Integrated Management of Rivers and Reservoirs

In Focus – Special Book Series

Integrated Management of Rivers and Reservoirs

Editors

Weiwei Shao and Quan Jin

IWA PUBLISHING

Published by **IWA Publishing**
Republic – Export Building, 1st Floor
2 Clove Crescent
London E14 2BE, UK
Telephone: +44 (0)20 7654 5500
Fax: +44 (0)20 7654 5555
Email: publications@iwap.co.uk
Web: www.iwapublishing.com

First published 2021

British Library Cataloguing in Publication Data
A CIP catalogue record for this book is available from the British Library

ISBN: 9781789062663 (paperback)

Contents

Editorial: Integrated management of rivers and reservoirs

PROMISING FEATURES

Climate change has altered hydrologic conditions, which directly affects water resources. For example, under the influence of rising temperatures, water vapor content in the atmosphere increases and precipitation changes, leading to a significant increase in the frequency and intensity of extreme hydrometeorological hazard events such as floods and droughts, and causing changes in water resources (Allen *et al.* 2010). In addition, increasing human activities have an impact on water resources, water environment and water ecology. Recent studies have shown that large-scale human activities in many areas have strongly influenced surface water balances (Farley *et al.* 2005; Murray *et al.* 2012). Rivers are important channels of water cycle, the main source of fresh water for human consumption and an important renewed resource, which play essential roles in material cycle, energy flow and climate regulation. Reservoirs are semi-natural and semi-artificial ecosystems, which are 'artificial lakes' formed by damming natural rivers (Hashimoto *et al.* 1982). They are important components of water conservancy constructions, and play vital roles in regional flood interception, water storage, and the regulation of water flow (Jackson *et al.* 1980; Labadie 2004). Rivers and reservoirs are natural and artificial water resource carriers. As a result, to promote ecological environment protection and high-quality development, it is of great significance to study the mechanism of the water cycle under the changing environment and build an optimized water resource management based on it.

However, rivers and reservoirs are faced with many problems, which have attracted the attention of experts and scholars. (i) Human activities and natural changes in the upper reaches of the reservoir, such as soil surface erosion caused by excessive agricultural activities, have led to rivers carrying large amounts of sediment into the lakes and reservoirs. (ii) The decrease in water inflow and increase in salinity of rivers, which are caused by climate change, excessive agricultural activities and inappropriate water withdrawal by human activities, have led to serious ecological disasters. (iii) Chemicals and pesticides have polluted rivers and reservoirs. It includes non-point source pollution caused by unconscionable cultivation and farming, as well as point source pollution caused by inadequate water sewage disposal in urban sewage treatment plants and direct discharge of domestic or industrial sewage. (iv) Nutrients, carried by domestic or industrial sewage, agricultural drainage and precipitation, flowing into reservoirs, as well as the change of flow velocity and the excess of nutrients, resulting in eutrophication. (v) The reservoirs are acidified due to the acid rain. (vi) Reservoirs have serious impacts on the river, which mainly include the shrinkage of river channels and the reduction of flood plains; destruction of aquatic habitats, with fewer birds and species; reduction or even disappearance of migratory fish; invasion of seawater and deterioration of estuarine ecosystems; and invasions of alien species (Kingsford 2000; Poff & Hart 2002). In view of the above problems, it is necessary to study the optimal management of water resources in rivers, lakes and reservoirs, the hydrological mechanism and ecological environment process (including experimental and numerical simulation methods), the optimal operation of engineering facilities and their natural and social effects, as well as the mitigation of water security problems and ensuring the sustainability of ecosystem through the integrated management of rivers and reservoirs (Richter *et al.* 2003, 2007).

THIS SPECIAL ISSUE

The papers in this special issue of *Water Supply* contribute to various aspects of integrated management of rivers and

doi: 10.2166/ws.2021.039

reservoirs. Topics covered include water quality related to environmental ecology and agriculture, reservoir management, and river engineering and floods.

This special issue compiles the recent main results of the integrated management of rivers and reservoirs. It confirms the modernization of reservoir management under the concept of coordinated development (Yu *et al.* 2020), and flood prevention emergency plans during reservoir construction based on generalized intuitionistic fuzzy soft sets and TOPSIS (Wu *et al.* 2020). Numerical simulation of groundwater pollutant transport in unsaturated flow (Yang *et al.* 2020) and an integrated allocation model of water quantity and quality in the Yellow River are explored (Zhou 2020). According to the impacts of precipitation and topographic conditions on the model simulation, a study suggests that the method of incorporating additional runoff generation module in the traditional model can significantly improve the accuracy of flood simulation (Sun *et al.* 2020). In addition, water security is evaluated based on the capacity for socio-economic regulation, which shows that strengthening economic and social control measures are conducive to solving the problems of regional water resource shortage (Tang *et al.* 2020).

Guest Editors

Weiwei Shao
State Key Laboratory of Simulation and Regulation of Water Cycle in River Basin, China Institute of Water Resources and Hydropower Research, Beijing, China

Quan Jin
China Institute of Water Resources and Hydropower Research, Beijing, China

REFERENCES

Allen, C. D., Macalady, A. K., Chenchouni, H., Bachelet, D., Mcdowell, N., Vennetier, M., Kitzberger, T., Rigling, A., Breshears, D. D. & Hogg, E. H. 2010 A global overview of drought and heat-induced tree mortality reveals emerging climate change risks for forests. *Forest Ecology and Management* **259** (4), 660–684.

Farley, K. A., Jobbagy, E. G. & Jackson, R. B. 2005 Effects of afforestation on water yield: a global synthesis with implications for policy. *Global Change Biology* **11** (10), 1565–1576. https://doi.org/10.1111/j.1365-2486.2005.01011.

Hashimoto, T., Stedinger, J. R. & Loucks, D. P. J. W. R. R. 1982 Reliability, resiliency, and vulnerability criteria for water resource system performance evaluation. *Water Resources Research* **18** (1), 14–20.

Jackson, T. A., Hecky, R. E. J. C. J. o. F., & Sciences, A. 1980 Depression of primary productivity by humic matter in lake and reservoir waters of the boreal forest zone. *Canadian Journal of Fisheries & Aquatic Sciences* **37** (12), 2300–2317.

Kingsford, R. T. J. A. E. 2000 Ecological impacts of dams, water diversions and river management on floodplain wetlands in Australia. *Austral Ecology* **25** (2), 109–127.

Labadie, J. W. 2004 Optimal operation of multireservoir systems: state-of-the-art review. *Journal of Water Resources Planning & Management* **130** (2), 93–111.

Murray, S. J., Foster, P. N. & Prentice, I. C. 2012 Future global water resources with respect to climate change and water withdrawals as estimated by a dynamic global vegetation model. *Journal of Hydrology* **448**, 14–29. https://doi.org/10.1016/j.jhydrol.2012.02.044.

Poff, N. L. & Hart, D. D. 2002 How dams vary and why it matters for the emerging science of dam removal. *Bioscience* **52** (8), 659–668.

Richter, B. D., Mathews, R. & Wigington, H. R. J. E. A. 2003 Ecologically sustainable water management: managing river flows for ecological integrity. *Ecological Applications* **13** (1), 206–224.

Richter, B. D., Thomas, G. A. J. E. & Society 2007 Restoring environmental flows by modifying dam operations. *Ecology and Society* **12** (1), 181–194.

Sun, L., Li, Z., Zhang, K. & Jiang, T. 2020 Impacts of precipitation and topographic conditions on the model simulation in the north of China. *Water Supply* **20** (8), ws2020284.

Tang, L., Zhang, W., Liu, Z. & Qi, Y. 2020 Evaluation of water security based on capacity for socio-economic regulation. *Water Supply* **20** (8), ws2020287.

Wu, H., Wang, J., Feng, J., Liu, D. & Liu, S. 2020 Research on the decision-making of flood prevention emergency plans during reservoir construction based on generalized intuitionistic fuzzy soft sets and TOPSIS. *Water Supply* **20** (8), 3665–3675.

Yang, J., Tao, Y., Ren, W. & Cao, P. 2020 Numerical simulation of groundwater contaminant transport in unsaturated flow. *Water Supply* **20** (8), 3730–3738.

Yu, D. J., Zou, Z. & Li, J. 2020 Evaluation of reservoir management modernization under the concept of coordinated development. *Water Supply* **20** (8), 3697–3706.

Zhou, K. 2020 Integrated allocation model of water quantity and quality and its application in the Yellow River. *Water Supply* **20** (8), 3768–3778.

A study on the phytoplankton community structure in the Diaohe River section of the Middle Route of the South-to-North Water Diversion Project in winter

Yi Xu, Jiankui Liang, Junqiang Lin, Xiaohui Lei, Gongbo Ding and Ruifang Yuan

ABSTRACT

As the first span and aqueduct of the Middle Route of the South-to-North Water Diversion Project, Diaohe Aqueduct has brought good engineering and social benefits since its completion. In order to understand the present water quality of the Diaohe River section of the route, four samplings were conducted at Diaohe Aqueduct in the winter of 2018. The main results of this study are as follows: 62 species (or variants) of phytoplankton belonging to 32 genera, 22 families and six phyla were identified during the study period, among which Bacillariophyta are the most dominant species. The identified phytoplankton fall into eight indicator grades: os; ps, αm; ps, αm, βm; αm; αm, βm; αm, βm, os; βm; and βm, os. A total of 17 functional groups are detected, among which TB is the absolute dominant functional group. Based on the composition characteristics of phytoplankton, the water quality trophic state of Diaohe Aqueduct is mesotrophic. However, other physical indexes show that the water quality of the Middle Route of the South-to-North Water Diversion Project is good. The conclusions drawn using phytoplankton alone as the evaluation index may not be comprehensive. In future research, how to build a comprehensive index system combining other water quality indexes for the Middle Route of the South-to-North Water Diversion Project needs yet to be improved and tested.

Key words | community structure, Diaohe Aqueduct, functional groups, phytoplankton, water quality

Yi Xu
College of Civil Engineering and Architecture,
Zhejiang University,
Hangzhou, Zhejiang Province, 310058,
China

Jiankui Liang
Construction and Administration Bureau of South-to-North Water Diversion Middle Route Project,
Beijing,100038,
China

Yi Xu
Junqiang Lin (corresponding author)
Xiaohui Lei
State Key Laboratory of Simulation and Regulation of Water Cycle in River Basin,
China Institute of Water Resources and Hydropower Research,
Beijing, 100038,
China
E-mail: *linjq@iwhr.com*

Gongbo Ding
State Key Laboratory of Hydraulics and Mountain River Engineering,
Sichuan University,
Chengdu, Sichuan Province, 610065,
China

Ruifang Yuan
College of Water Resources and Environment,
China University of Geosciences (Beijing),
Beijing, 100083,
China

HIGHLIGHTS

- Sixty-two species of phytoplankton were measured in Diaohe Aqueduct of MPR in winter.
- A diatom, such as *Fragilaria*, *Cymbella*, was the main dominant species.
- TB functional group is the absolute dominant functional group.
- The water quality trophic state of the Diaohe Aqueduct is mesotrophic.
- The water quality in the Diaohe Aqueduct during the study period is clean.

INTRODUCTION

As the largest inter-basin water diversion project in the world, the Middle Route of the South-to-North Water Diversion Project (MRP) is a major strategic infrastructure for alleviating water shortages in northern China, reasonably allocating water resources, ensuring sustainable economic and social development, and comprehensively building a well-off society in China. Since the MRP provides domestic and industrial water to northern China (e.g. Beijing), the water quality in the MRP is required to meet the Grade II Chinese Environmental Quality Standard for Surface

doi: 10.2166/ws.2020.247

Water (GB3838-2002; Tang *et al.* 2011). However, the MRP is composed of a long canal and complex hydraulic structures, and transfers drinking water to Beijing, Tianjin and other cities in open channels, which has caused a certain threat to the MRP water quality (Tang *et al.* 2016; Zhu *et al.* 2019). Therefore, it is particularly important to monitor the water quality in the MRP.

Phytoplankton is the principal primary producer in aquatic ecosystems, it is highly sensitive and responds rapidly to environmental changes (Yuan *et al.* 2018), and the changes of its community structure have a direct or indirect impact on other aquatic organisms, and can even affect the stability of the aquatic ecosystem. Learning phytoplankton structure could obtain a comprehensive and real understanding of the status of an aquatic ecosystem (Suikkanen *et al.* 2007). Many researchers determine the water quality of a study area by studying phytoplankton communities and environmental variables, for example, Gharib *et al.* (2011) determined the water quality of Matru Beach in the Mediterranean Sea in southeastern Egypt by studying environmental variables and the abundance and community structure of phytoplankton during 2009–2010. Zaghloul *et al.* (2020) studied the species composition and abundance of phytoplankton weekly from 16 June 2011 to 30 May 2012 in the Eastern Harbor, and divided the aquatic system into oligotrophic, mesotrophic and eutrophic water types. However, due to the short running time of the MPR, there are few studies on determining water quality by studying phytoplankton community. Therefore, it is necessary to monitor the water quality in the MRP by studying the phytoplankton community.

Known as 'the first span of the massive water diversion project', Diaohe Aqueduct is the first aqueduct for water conveyance in the MRP, and it has brought good engineering and social benefits since its completion. As the beginning section of the Middle Route of the South-to-North Water Diversion Project, the water quality of the Diaohe River section has a direct impact on the condition of drinking water for residents in the following areas receiving the water supply. Therefore, this paper investigates the phytoplankton in the Diaohe River section of the Middle Route of the South-to-North Water Diversion Project, with the aim of providing an important basis for the ecological protection of the water of this main canal (Diaohe River section) in the Middle Route.

MATERIALS AND METHODS

Study area

The beam-type Diaohe Aqueduct is located on the river 500 m away from Yaoying Village, Jiulong Town, Dengzhou City, Henan Province, 2.5 km south from Dengzhou–Zhechuan Highway. The main structures of the aqueduct include 115.1 m of surplus escape, 40 m of inlet transition, 26 m of inlet controlling sluice, 20 m of inlet connection, 350 m of aqueduct, 20 m of outlet connection, 15 m of outlet maintenance chamber, 70 m of outlet transition, and 3.9 m of outlet open channel, with the total length being 660 m. The span arrangement of the aqueduct body is 8*40 m + 1*30 m, the upper structure of which is a pre-stressed box-type simply supported beam arranged in double lines and double troughs. The net width of a single trough is 13.0 m, with a maximum cross-sectional area of 83.98 m^2, a designed water depth of 5.75 m, and an increased water depth of 6.46 m. The trough is 15 m wide at the top, 15.1 m at the bottom, the inner wall spacing between the two troughs is 5.0 m, and a sidewalk slab is laid between them. The full width of the double-line aqueduct is 33.0 m wide at the top and 33.1 m wide at the bottom. Diaohe Aqueduct is the first water conveyance aqueduct on the main canal of the Middle Route of the South-to-North Water Diversion Project, known as 'the first span of the massive water diversion project'. From December 12, 2014, when the Middle Route of the South-to-North Water Diversion Project was officially put into use, to October 1, 2017, the total cumulative amount of water entering the canal has reached 10 billion m^3. The aqueduct has yielded good engineering and social benefits.

Sampling time and sites

In this experiment, four samplings were conducted at the Middle Route (Diaohe section) of the South-to-North Water Diversion Project in November 2018 (on November 7, 14, 20, and 27, 2018, respectively). The monitored canal section (Diaohe section) is in two different channel shapes: straight and curved. The light intensity varies on different sides of it, and the curved section has concave and convex banks which lead to different flow velocities.

All these factors will affect the growth and distribution of algae. As a result, three sampling points were set up after comprehensively considering the above circumstances and the operability on the site. They are located on the left and right banks of the pontoon bridge of Diaohe Aqueduct (straight section) and the water quality sensitive position of Diaohe Aqueduct (curved section), as shown in Figure 1.

Sample collection and analyses

Environmental factor analyses

The pH, water temperature (WT), dissolved oxygen (DO), and conductivity, oxidation–reduction potential (ORP) were measured by YSI 6600EDS at 0.5 m below the water surface. At the same, a deep water sampler was used to collect the mixed water sample for an additional index at 0.5 m below the water surface, and 500 ml of mixed water was extracted and put into a polyethylene plastic bottle, then the water samples were transported to the laboratory within 24 h, and stored in a refrigerator at 4 °C before being analyzed.

Additional indexes included the total phosphorus, orthophosphate, total nitrogen, ammonia nitrogen, nitrate nitrogen, permanganate index, chemical oxygen demand and chlorophyll *a*. These indexes were measured according to Chinese standard methods (see Wei *et al.* (2002) for details).

Algae sample collection and identification

Collection of samples for qualitative analysis: use the plankton net no. 25 (mesh diameter being 0.064 mm) to collect the sample. Specifically, drag the net slowly at a speed of 20–30 cm/s by making a '∞' shape repetitively for 3–5 min in the space between the water surface and 0.5 m deep below the water surface. Add Lugol's solution into the collected sample for fixation, label the sample bottle, and make various records (Jin & Tu 1990). To perform qualitative analysis of the phytoplankton, classify the plankton samples, make them into temporary specimens for microscope slides and then identify them. In this process, the phytoplankton need to be photographed and preserved, and identified by authoritative books (Shi 2004; Zhou & Chen 2004; Wang *et al.* 2008).

Collection of samples for quantitative analysis: use the plexiglass water sampler to collect the sample by holding it still at about 0.5 m below the water surface, collect 1 L of water, put it into the sampling bottle that has been soaked with nitric acid and cleaned beforehand, and then immediately add 15 ml of Lugo's solution with a mass fraction of 15‰ into the collected sample for fixation. After taking the sample back to the laboratory, shake it thoroughly and transfer it into a 1,000 ml separatory funnel. After leaving the separatory funnel still for 24 h, aspirate the supernatant from it by siphoning (using a siphon) until 250 ml of the sample is left, and remove the remaining liquid to a small container. Leave the sample still for 24 h again, and then pipette the supernatant until 30 ml of the sample is left. After shaking the remaining liquid thoroughly, pipette 0.1 ml of the shaken sample into a phytoplankton counting box (0.1 ml) and count the phytoplankton under a microscope with a 10 × 40 magnification lens (Liu *et al.* 2019). The cell count on each slide is no less than 200, and the average of the counts of three slides is taken for each specimen. Since the specific gravity of phytoplankton is close to 1, the volume of phytoplankton can be directly converted to biomass (wet weight). That is, the biomass of phytoplankton is equal to the number of phytoplankton multiplied by their average volume, and the unit is mg/L. The biomass of unicellular organisms is mainly obtained by measuring their individual shape (Jin & Tu 1990).

Statistical analysis

The following equations of three diversity indices are used to calculate community metrics of phytoplankton. These are the Shannon–Wiener diversity index (Shannon & Weaver 1949) (shown in Equation (1)), Margalef richness index (Margalef 1958) (shown in Equation (2)), the Pielou evenness index (Pielou 1969) (shown in Equation (3)):

$$H' = -\sum_{i=1}^{S} p_i \ln p_i \tag{1}$$

$$d = 1 - \sum_{i=1}^{S} p_i^2 \tag{2}$$

$$J = \frac{H'}{\ln S} \tag{3}$$

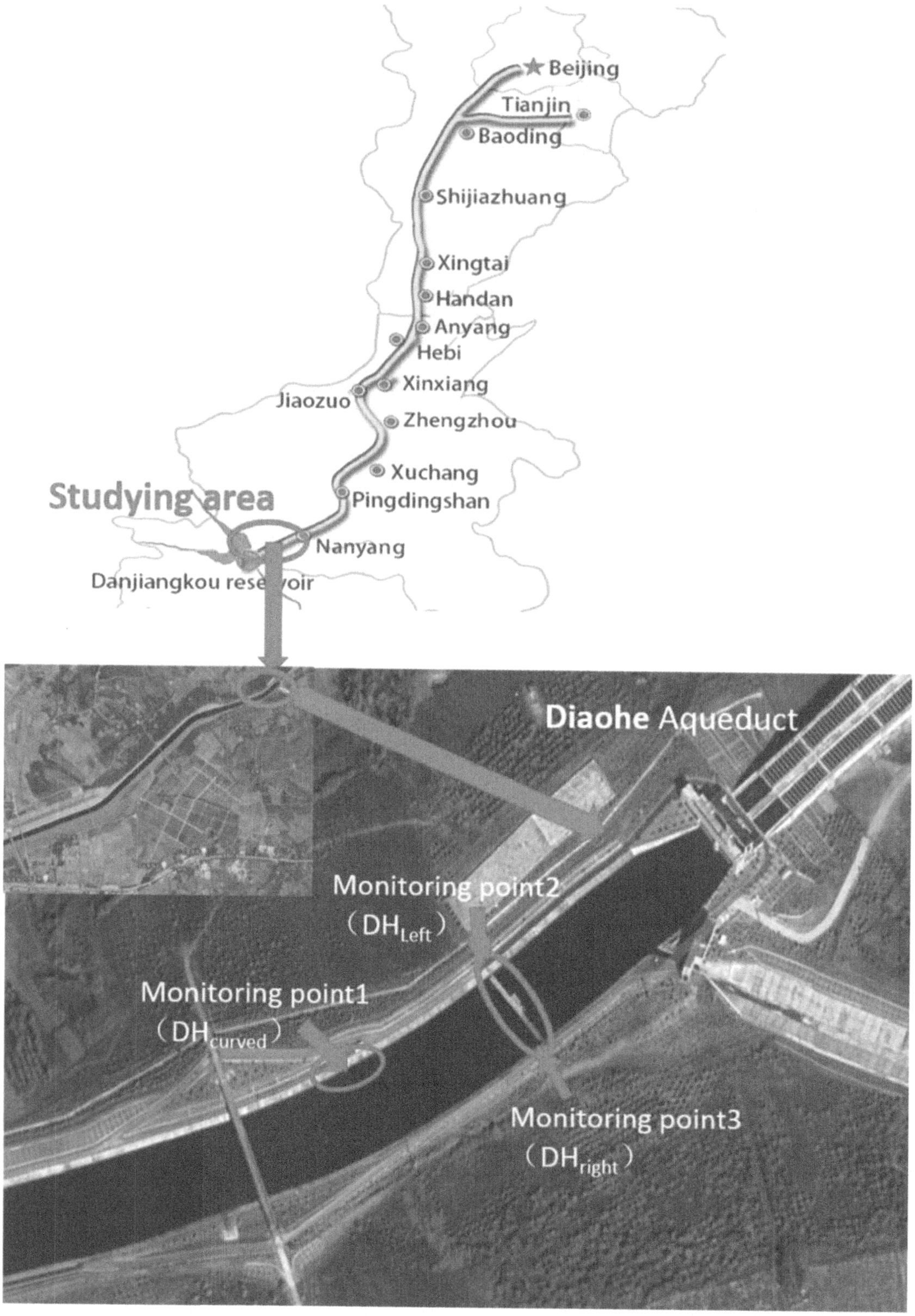

Figure 1 | Algae sampling points at Diaohe Aqueduct.

where H' is the Shannon–Wiener index, d is the Margalef richness index, J is the Pielou evenness index, p_i is the ratio of the individual number of the species i and total individual number (N), and S is the number of species within the given sample; and Excel is used in the following statistical analyses.

RESULTS AND DISCUSSION

Water quality indexes

The water quality indexes in Diaohe Aqueduct during the study period are summarized in Table 1. During the study period, the water temperature was maintained at about 14 °C, and pH was maintained at about 8.5. Chlorophyll *a* is an index that characterizes the content of algae; comparing the content of chlorophyll *a* at different times in the same place, it is found that as time goes by, the content of chlorophyll *a* shows a downward trend, while the content of nutrients and organic matter both show a downward trend, and the water temperature does not change. It is inferred that the phytoplankton had begun to enter a dormant period. Comparing with the Chinese Environmental Quality Standard for Surface Water (G3838-2002), it is found that the content of all water quality indexes meet the Grade II Chinese Environmental Quality Standard for Surface Water, that means the water quality in the Diaohe Aqueduct during the study period is clean.

Composition of phytoplankton community

In the samples collected at Diaohe Aqueduct in November 2018, Chlorophyta, Cyanophyta, Bacillariophyta, Pyrrophyta, Euglenophyta and Cryptophyta are identified. There are 62 species (or variants) of phytoplankton under 32 genera, 22 families and six phyla, including 30 species (or variants) of Bacillariophyta under 12 genera, accounting for 48.39% of the total phytoplankton; 16 species of Chlorophyta under 11 genera, accounting for 25.81%; nine species of Cyanophyta under four genera, accounting for 14.52%; two species of Pyrrophyta under two genera, accounting for 3.23%; one species of Euglenophyta under one genus, accounting for 1.61%; and four species of Cryptophyta under two genera, accounting for 6.45% (Figure 2 shows the composition of phytoplankton and Table 2 lists the phytoplankton species during the study period).

Composition of phytoplankton pollution indicator species

As early as 1908, German scientists Kolkwitz and Marsson (Kolkwitz & Marsson 1908) used algae to monitor the environmental quality of water bodies, and they were the first to propose a sewage biological system indicating

Table 1 | Water quality indexes in the Diaohe Aqueduct during the study period (except for WT and pH, all indexes are in mg/L)

Sampling time	Monitoring point	pH	WT	DO	COD_{cr}	I_{Mn}	TP	PO_4^{3-}-P	NH_4^+-N	NO_3^--N	Chla
2018/11/6	DH_{left}	8.3	14.4	8.1	13.54	1.75	5.75×10^{-3}	3.47×10^{-3}	0.028	0.657	2.14
2018/11/6	DH_{right}	8.5	15.3	8.08	10.53	1.62	5.25×10^{-3}	3.10×10^{-3}	0.031	0.495	2.10
2018/11/6	DH_{curved}	8.4	14.6	8.1	12.04	1.51	5.47×10^{-3}	2.87×10^{-3}	0.022	0.627	3.56
2018/11/13	DH_{left}	8.5	13.8	8.24	12.36	2.76	5.47×10^{-3}	2.65×10^{-3}	0.04	0.582	1.40
2018/11/13	DH_{right}	8.5	14.0	8.2	11.22	2.08	6.07×10^{-3}	3.06×10^{-3}	0.032	0.545	2.14
2018/11/13	DH_{curved}	8.6	14.4	8.26	11.60	2.17	4.42×10^{-3}	2.60×10^{-3}	0.014	0.539	1.84
2018/11/20	DH_{left}	8.4	13.5	8.1	11.20	1.29	6.16×10^{-3}	3.47×10^{-3}	0.016	0.545	1.54
2018/11/20	DH_{right}	8.4	14.3	8.08	12.60	1.20	5.79×10^{-3}	4.56×10^{-3}	0.029	0.506	2.13
2018/11/20	DH_{curved}	8.4	12.5	8.13	13.21	1.63	4.93×10^{-3}	3.88×10^{-3}	0.029	0.552	1.20
2018/11/27	DH_{left}	8.1	14.1	8.08	9.67	1.58	5.79×10^{-3}	3.06×10^{-3}	0.017	0.496	2.13
2018/11/27	DH_{right}	8.5	13.7	8.1	9.15	1.55	7.25×10^{-3}	4.84×10^{-3}	0.028	0.371	1.93
2018/11/27	DH_{curved}	8.2	14	8.11	9.57	1.58	6.34×10^{-3}	3.56×10^{-3}	0.017	0.513	2.25

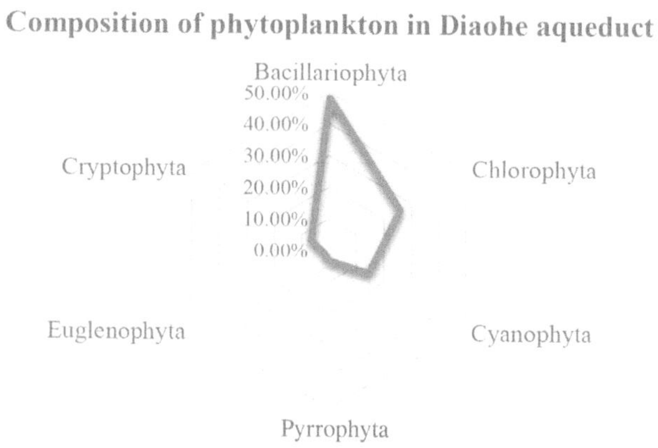

Figure 2 | Composition of phytoplankton in the Diaohe Aqueduct.

organic pollution in rivers. It pointed out different types of 'indicator organisms' for different pollution zones, and these were low pollution zone, β medium pollution zone, α medium pollution zone and multi-contamination zone. Through the changes of 'indicating organisms' in different water bodies, the water quality of the water body can be judged. Since the concept of 'indicating organisms' was proposed, it has been used by many scholars and has been continually revised and improved.

In the samples collected at Diaohe Aqueduct in November 2018, 62 species (or variants) of phytoplankton under 32 genera, 22 families and six phyla are identified, and they fall into eight indicator grades: os (oligotrophic); ps (p hypertrophic), αm (α medium eutrophic); ps, αm, βm (β mesotrophic); αm; αm, βm; αm, βm, os; βm; and βm, ps. Among them, there are one species of os phytoplankton under one genus, accounting for 5.56% of the total indicator species; two species of αm phytoplankton under two genera, accounting for 11.11%; four species of βm phytoplankton under four genera, accounting for 22.22%; four genera of βm, ps phytoplankton under three genera, accounting for 22.22%; and two species of α-β-o mesotrophic phytoplankton under two genera, accounting for 11.11%. The phytoplankton pollution indicator species and their quantity in Diaohe Aqueduct are shown in Figure 3.

In this experiment, the identified pollution indicator species fall into six phyla, namely Chlorophyta, Cyanophyta, Bacillariophyta, Pyrrophyta, Euglenophyta and Cryptophyta. The Bacillariophyta include *Cyclotella meneghiniana* Kütz., *Gomphonema* sp., *Melosira granulata* (Ehr.) Ralfs, and *Synedra ulna* (Nitzsch) Ehr.; the *Pyrrophyta* include *Ceratium* sp.; the Cyanophyta include *Microcystis robusta* (Clark) Nygaard, and *Aphanizomenon flos-aquae* (L.) Ralfs; the Euglenophyta include *Euglena viridis* Ehr.; the Chlorophyta include *Phacotus lenticularis* (Ehr.) Stein, *Eudorina* sp., *Oocystis* sp., *Pediastrum simplex* Meyen, *Planktosphaëria gelatinosa* G. M. Smith, *Radiococcus nimbatus* (Wild.) Schm., *Pediastrum duplex* Meyen, and *Staurastrum manfeldtii* Delp.; and the Cryptophyta include *Cryptomonas erosa* Ehr., as shown in Table 3.

Phytoplankton diversity

The Shannon–Wiener diversity index (H') is a single-factor index for representing the relationship between the number and species of biomes with multiple kinds of organisms. The biodiversity in communities reflects the complexity of biomes or habitats. This index is widely used to monitor changes of marine benthos (Noman *et al.* 2019), and it also can be used as a main indicator to evaluate water quality. As shown in Table 4 (Table 4 shows water quality and ecological evaluation standards), the Shannon–Wiener diversity index (H'), the Margalef richness index (d) and the Pielou evenness index (J) were applied to evaluate the water quality and ecological environment of Diaohe Aqueduct (Figure 4 shows the changes in the biodiversity index for each sampling points during the study period).

Water quality evaluation with the Shannon–Wiener diversity index

The Shannon–Wiener diversity index of phytoplankton in Diaohe Aqueduct ranges from 1.42 to 1.66. The biggest value 1.66 was recorded at the curved section, followed by 1.56 at the monitoring point on the left bank, and the smallest value 1.42 on the right bank. The range of changes is relatively small, and the pollution level is rated medium (Miao *et al.* 2019).

Water quality evaluation with the Margalef richness index

The Margalef richness index indicates the richness of species in a biological community. Generally speaking, the

Table 2 | List of phytoplankton species in the Diaohe Aqueduct

Phyla	Genera	Species
Chlorophyta	*Pediastrum* Meyen	*Pediastrum simplex* Meyen
Chlorophyta	*Pediastrum* Meyen	*Pediastrum duplex* Meyen
Chlorophyta	*Planktosphaëria*	*Planktosphaëria gelatinosa* G.M. Smith
Chlorophyta	*Staurastrum* Meyen	*Staurastrum planctonicum* Teiling
Chlorophyta	*Staurastrum* Meyen	*Staurastrum planctonicum*
Chlorophyta	*Radiococcus* Schmidle	*Radiococcus nimbatus* (Wild.) Schm.
Chlorophyta	*Pediastrum* Meyen	*Pediastrum clathratum* (Schroeter) Lemmermann
Chlorophyta	*Phacotus*	*Phacotus lenticularis* (Ehr.) Stein
Chlorophyta	*Eudorina*	*Eudorina* sp.
Chlorophyta	*Oocystaceae*	*Oocystis* sp.
Chlorophyta	*Staurastrum* Meyen	*Staurastrum manfeldtii* Delp.
Chlorophyta	*Scenedesmus* Meyen	*Scenedesmus bijuga* (Turp.) Lag.
Chlorophyta	*Coelastrum* Nägeli	*Coelastrum reticulatum*
Chlorophyta	*Chlamydomonas* Ehrenherg	*Chlamydomonas* sp.
Chlorophyta	*Scenedesmus* Meyen	*Scenedesmus* sp.
Chlorophyta	*Mougeotia* Meyen	*Mougeotia* sp.
Cyanophyta	*Oscillatoria* Vaucher	*Oscillatoria tenuis* Ag.
Cyanophyta	*Oscillatoria* Vaucher	*Oscillatoria* sp.
Cyanophyta	*Anabaena* Bory	*Anabaena* sp.
Cyanophyta	*Aphanizomenon* Morr	*Aphanizomenon* sp.
Cyanophyta	*Aphanizomenon* Morr	*Aphanizomenon flos-aquae* (L.) Ralfs
Cyanophyta	*Microcystis* Kütz.	*Microcystis* sp.
Cyanophyta	*Microcystis* Kütz.	*Microcystis densa* G.S.West
Cyanophyta	*Microcystis* Kütz.	*Microcystis ichthyoblabe* Kütz.
Cyanophyta	*Microcystis* Kütz.	*Microcystis robusta* (Clark) Nygaard
Bacillariophyta	*Gyrosigma* Hassal	*Gyrosigma* sp.
Bacillariophyta	*Gyrosigma* Hassal	*Gyrosigma acuminatum* (Kütz.) Raben.
Bacillariophyta	*Fragilaria* Lyngbye	*Fragilaria* sp.
Bacillariophyta	*Fragilaria* Lyngbye	*Fragilaria capucina* Desmaziéres
Bacillariophyta	*Fragilaria* Lyngbye	*Fragilaria intermedia* (Grun.) Grun.
Bacillariophyta	*Diatoma* De Candolle	*Diatoma vulgare* Bory
Bacillariophyta	*Diatoma* De Candolle	*Synedra tenue* Agardh
Bacillariophyta	*Nitzschia* Hassall	*Hantzschia* sp.
Bacillariophyta	*Nitzschia* Hassall	*Nitzschia sigmoidea* (Nitzsch.) Wm. Smith
Bacillariophyta	*Cymbella* Agardh	*Cymbella* sp.
Bacillariophyta	*Cymbella* Agardh	*Cymbella aspera*
Bacillariophyta	*Cymbella* Agardh	*Cymbella affinis* Kuetz
Bacillariophyta	*Achnanthes* Bory	*Achnanthes* sp.
Bacillariophyta	*Cyclotella* Kützing	*Cyclotella* sp.
Bacillariophyta	*Cyclotella* Kützing	*Cyclotella hubeiana* Chen et Zhu

(continued)

Table 2 | continued

Phyla	Genera	Species
Bacillariophyta	*Cyclotella* Kützing	*Cyclotella meneghiniana* Kütz.
Bacillariophyta	*Asterionella* Hassall	*Asterionella* sp.
Bacillariophyta	*Gomphonema* Ehrenberg	*Gomphonema* sp.
Bacillariophyta	*Gomphonema* Ehrenberg	*Gomphonema subclavatum*
Bacillariophyta	*Synedra* Ehrenberg	*Synedra* sp.
Bacillariophyta	*Synedra* Ehrenberg	*Synedra acus* Kütz.
Bacillariophyta	*Synedra* Ehrenberg	*Synedra ulna* (Nitzsch) Ehr.
Bacillariophyta	*Melosira* Agardh	*Melosira varians* Agardh
Bacillariophyta	*Melosira* Agardh	*Melosira granulata* (Ehr.) Ralfs
Bacillariophyta	*Melosira* Agardh	*Melosira granulata* var. *angustissima* Müll.
Bacillariophyta	*Melosira* Agardh	*Melosira granulata* var. *angustissima* f. *spiralis* Hustedt
Bacillariophyta	*Navicula* Bory	*Navicula* sp.
Bacillariophyta	*Navicula* Bory	*Navicula radiosa* Kuetz
Bacillariophyta	*Navicula* Bory	*Navicula cincta* (Ehr.) Ralfs
Bacillariophyta	*Navicula* Bory	*Navenacea*
Pyrrophyta	*Certium* Schr.	*Ceratium* sp.
Pyrrophyta	*Gymnodinium* Stein	*Gymnodinium* sp.
Euglenophyta	*Euglena* Ehrenberg	*Euglena viridis* Ehr.
Cryptophyta	*Cryptomonas* Ehr.	*Cryptomonas ovata* Ehr.
Cryptophyta	*Cryptomonas* Ehr.	*Cryptomonas erosa* Ehr.
Cryptophyta	*Chroomonas*	*Chroomonas acuta* Uterm.
Cryptophyta	*Chroomonas*	*Chroomonas caudata* Geitler

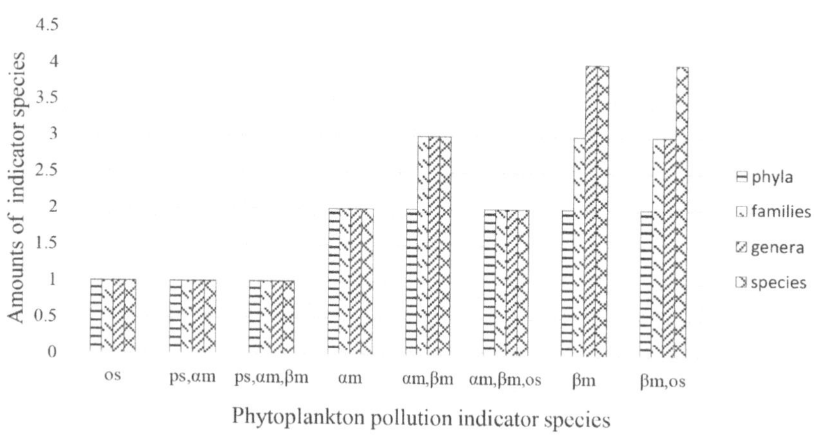

Figure 3 | Amounts of phytoplankton pollution indicator species.

number of species is high in a healthy environment, and the number of species is low in a polluted environment (Ma 2012).

The Margalef richness index of phytoplankton in Diaohe Aqueduct ranges from 6.98 to 7.21. The range of changes is big, with the characteristics of index changes in

Table 3 | Composition of phytoplankton pollution indicator species

Phytoplankton pollution indicator species	Monitoring point			Indicator grade
	DH_{left}	DH_{right}	DH_{curved}	
Synedra ulna (Nitzsch) Ehr.	+	+	+	os
Gomphonema sp.	+	+	+	αm
Cyclotella meneghiniana Kütz.		+		αm, βm, os
Melosira granulata (Ehr.) Ralfs	+	+	+	βm
Ceratium sp.			+	βm, os
Aphanizomenon flos-aquae (L.) Ralfs	+	+	+	αm, βm
Microcystis robusta (Clark) Nygaard	+	+	+	αm, βm
Euglena viridis Ehr.		+		ps, αm
Phacotus lenticularis (Ehr.) Stein	+			αm
Eudorina sp.			+	αm, βm
Oocystis sp.			+	αm, βm, os
Pediastrum simplex Meyen		+		βm
Planktosphaëria gelatinosa G.M.Smith	+			βm
Radiococcus nimbatus (Wild.) Schm.	+			βm
Pediastrum duplex Meyen	+	+	+	βm, os
Staurastrum planctonicum Teiling	+			βm, os
Staurastrum manfeldtii Delp.		+	+	βm, os
Cryptomonas erosa Ehr.	+			ps, αm, βm

space being left bank> water quality sensitive point> right bank. The index at all the three monitoring points is greater than 6, so the water quality is relatively healthy and the pollution level is clean.

Water quality evaluation with the Pielou evenness index

The Pielou evenness index gives a good representation of the distribution uniformity of the individuals of phytoplankton

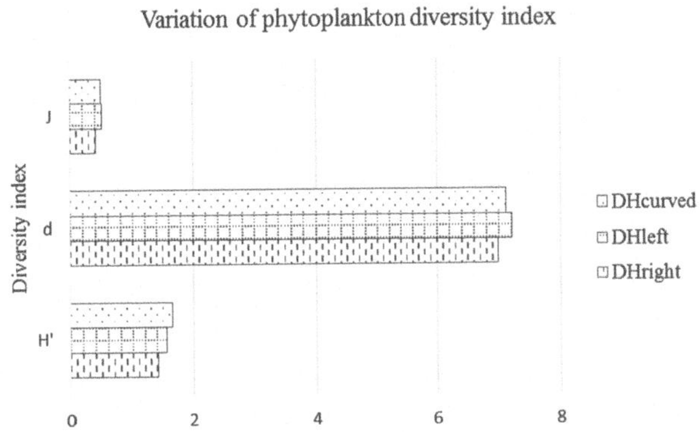

Figure 4 | Spatial variations of phytoplankton diversity indexes.

species. The value range of J is 0–1. When this value is big, it indicates that the distribution of individuals among the species is uniform; when this value is small, it suggests uneven distribution of individuals among the species (Ma 2012).

The Pielou evenness index of phytoplankton in Diaohe Aqueduct ranges from 0.42 to 0.51, and the changes are stable, so the pollution level is rated as medium-light pollution. The value at the monitoring point on the right bank is small, while that on the left bank is large.

The spatial variation of the Pielou evenness index of phytoplankton in Diaohe Aqueduct is characterized as water quality sensitive point > left bank > right bank. The water quality at the monitoring points on the left and right banks is worse than that at the water quality sensitive point. Their pollution level is medium pollution, with the water quality on the left bank better than that on the right bank and rated as light pollution. Generally speaking, the distribution of individuals among the phytoplankton species in Diaohe Aqueduct is uniform.

Classification of phytoplankton into functional groups

The composition of phytoplankton functional groups can be analyzed according to the physiological and growth

Table 4 | Water quality and ecological evaluation standards

Biodiversity index	Serious pollution	Heavy pollution	Medium pollution	Light pollution	Oligosaprobic (clean)
H'		<1	1–2	2–3	>3
D	0–1	1–2	2–3	3–4	>4
J		0–0.3	0.3–0.5	0.5–0.8	0.8–1

characteristics of phytoplankton and the mechanism of their environmental adaptability as proposed by Reynolds *et al.* (2002) and Padisák *et al.* (2009). The identified phytoplankton were classified into different groups according to their functions. As shown in Table 5, the phytoplankton in Diaohe Aqueduct can be divided into 17 functional groups, namely: B, C, D, F, G, H1, J, LO, M, MP, N, P, T, TB, W1, WO, and X2.

If the dominance of a certain functional group is ≥0.02, its status in the water is dominant and we call it the dominant functional group (Wang *et al.* 2018). The monitoring results on the left bank of Diaohe River register the most species of Bacillariophyta, in which *Fragilaria*, *Cymbella*, *Navicula* and *Diatoma vulgare* Bory (in the order of dominance from large to small, the same below) are absolutely dominant. On the right bank of Diaohe River, *Fragilaria*, *Navicula* and *Cymbella* are absolutely dominant, and at the water quality sensitive point, *Navicula*, *Fragilaria*, *Cymbella* and *Oscillatoria* are absolutely dominant.

In the monitoring period, the degree of dominance of the TB functional group in all the four sampling investigations is ≥0.02, which makes it the absolute dominant functional group in the Diaohe section of the middle water diversion route. The TB functional group includes *Fragilaria* and *Navicula*, and the secondary functional groups are P and MP, which mainly include *Diatoma*,

Table 5 | Classification of phytoplankton in Diaohe Aqueduct into functional groups

Functional group	Representative species (genus)	Physiological characteristics
B	*Cyclotella* Kützing	Small and medium-sized mesotrophic lakes, sensitive to stratification
C	*Cyclotella hubeiana* Chen et Zhu, *Cyclotella meneghiniana* Kütz., *Asterionella* sp.	Small and medium-sized mesotrophic lakes, sensitive to stratification
D	*Synedra ulna* (Nitzsch.) Ehr., *Synedra acus* Kütz.	Shallow turbid rivers
F	*Oocystis* Nägeli, *Planktosphaëria gelatinosa* G. M. Smith	Deep mesotrophic and eutrophic lakes, thermocline
G	*Eudorina* sp.	Small eutrophic lakes, river basins or reservoirs
H1	*Anabaena* sp., *Aphanizomenon* sp., *Aphanizomenon flos-aquae* (L.) Ralfs	Shallow eutrophic, stratified, low-nitrogen lakes
J	*Pediastrum* Meyen, *Pediastrum simplex* Meyen, *Scenedesmus* Mey., *Scenedesmus bijuga* (Turp.) Lag.	Mixed shallow, highly eutrophic waters
LO	*Radiococcus nimbatus* (Wild.) Schm., *Gymnodinium* sp., *Gyrosigma* Hassal	Broad adaptability
M	*Microcystis* Kütz., *Microcystis densa* G.S.West, *Microcystis robusta* (Clark) Nygaard	Stable mesotrophic and eutrophic waters with transparency not too low
MP	*Oscillatoria* Vauch., *Cymbella* Agardh	Frequently disturbed, non-organic, shallow turbid lakes
N	*Staurastrum planctonicum* Teiling, *Staurastrum manfeldtii* Delp.	Mixed layer of mesotrophic waters,
P	*Diatoma* De Candolle, *Melosira* Agardh	Shallow highly mixed mesotrophic and eutrophic waters
T	*Mougeotia* sp.	Thermocline of uniformly mixed deep waters
TB	*Navicula* Bory, *Fragilaria* Lyngb., *Nitzschia* Hassall, *Achnanthes* Bory	Abominable streams and rivers
W1	*Euglena* Ehrenberg	Waters rich in organic matter or agricultural wastewater and domestic sewage
WO	*Chlamydomonas* Ehrenherg	Rivers and ponds highly rich in organic matter
X2	*Cryptomonas* Ehr., *Phacotus*	Shallow highly mixed, mesotrophic and eutrophic waters

Cymbella and *Oscillatoria*. All the three functional groups are between mesotrophic and eutrophic. In this study, since the TB, P, and MP functional groups are the dominant functional groups in Diaohe Aqueduct in winter, it is inferred that the nutritional state of Diaohe Aqueduct may be mesotrophic.

Discussion

During the study period, although the β-p hypertrophic phytoplankton indicator and β mesotrophic phytoplankton indicator dominated in the Diaohe Aqueduct, considering the phytoplankton diversity index (including the Shannon–Wiener index, the Magalef richness index and the Pielou evenness index) and the study of phytoplankton functional groups, it can be inferred that the water quality of the Diaohe section was mesotrophic. The Middle Route of the South-to-North Water Diversion Project diverts water from Danjiangkou Reservoir to Beijing, Tianjin and other cities. As Wan *et al.* (2020) reports, the water quality of Danjiangkou Reservoir in 2018 was in a mesotrophic state, and the Diaohe Aqueduct is only a section of the MRP, so it is speculated that the mesotrophic state of water quality in the Diaohe Aqueduct during the study period may be related to the water quality of the Danjiangkou Reservoir. However, the conclusions drawn using phytoplankton alone as the evaluation index may not be comprehensive, and needs to be further combined with other indexes for comprehensive consideration. So in the future the water quality index and the phytoplankton need to be considered comprehensively.

CONCLUSION

By analyzing the water quality monitoring data, it is found that the water quality of the Diaohe Aqueduct is relatively clean, which meets the Grade II Chinese Environmental Quality Standard for Surface Water (G3838-2002). The phytoplankton population in the Diaohe section of the Middle Route in winter consists of 62 species of phytoplankton under 32 genera, 22 families and six phyla, and the main dominant species or genera are *Cyclotella*, *Navicula*, *Fragilaria*, *Diatoma* and *Cymbella* under Bacillariophyta and *Oscillatoria* under Cyanophyta. Considering the pollution indicator species, dominant functional groups and diversity index of phytoplankton in the Diaohe Aqueduct, it is inferred that the water quality of the Diaohe section was mesotrophic in the winter of 2018. Although the conclusions analyzed by phytoplankton community may not be comprehensive, it is recommended that the local government should pay more attention to changes in the phytoplankton community while carrying out long-term water quality monitoring.

ACKNOWLEDGEMENTS

This paper was supported by the Major Science and Technology Program for Water Pollution Control and Treatment (2017ZX07108-001) and Special Project on Water Quality Guarantee and Environmental Impact of the Middle Route of South-to-North Water Transfer Project (JZ0205A222018).

DATA AVAILABILITY STATEMENT

All relevant data are included in the paper or its Supplementary Information.

REFERENCES

Gharib, S. M., El-Sherif, Z. M., Abdel-Halim, A. M. & Radwan, A. A. 2011 Phytoplankton and environmental variables as a water quality indicator for the beaches at Matrouh, south-eastern Mediterranean Sea, Egypt: an assessment. *Oceanologia* **53**, 819–836.

Jin, X. & Tu, Q. 1990 *Standards for Investigation of Lake Eutrophication*, 2nd edn. China Environmental Science Press, Beijing, China.

Kolkwitz, R. & Marsson, M. 1908 Ökologie der pflanzlichen Saprobien. *Berichte der Deutschen Botanischen Gesellschaft* **26**, 505–519.

Liu, J., Chen, Y., Li, M., Liu, B., Liu, X., Wu, Z., Cai, Y., Xu, J. & Wang, J. 2019 Water-level fluctuations are key for phytoplankton taxonomic communities and functional groups in Poyang Lake. *Ecological Indicators* **104**, 470–478.

Ma, L. 2012 *Assessing the Water Quality by Means of an Index of Biotic Integrity for Phytoplankton in Tanjiang River*. Master's thesis, Hebei University, Baoding, China.

Margalef, D. R. 1958 Information theory in ecology. *International Journal of General Systems* **3**, 36–71.

Miao, X., Wang, S., Liu, M., Ma, J., Hu, J., Li, T. & Chen, L. 2019 Changes in the phytoplankton community structure of the Backshore Wetland of Expo Garden, Shanghai from 2009 to 2010. *Aquaculture and Fisheries* **4**, 198–204.

Noman, M. A., Sun, J., Gang, Q., Guo, C., Islam, M. S., Li, S. & Yue, J. 2019 Factors regulating the phytoplankton and tintinnid microzooplankton communities in the East China Sea. *Continental Shelf Research* **181**, 14–24.

Padisák, J., Crossetti, L. O. & Naselli-Flores, L. 2009 Use and misuse in the application of the phytoplankton functional classification: a critical review with updates. *Hydrobiologia* **621**, 1–19.

Pielou, E. C. 1969 *An Introduction to Mathematical Ecology*. Wiley Interscience, New York, USA, pp. 23–36.

Reynolds, C. S., Huszar, V., Kruk, C., Naselli-Flores, L. & Melo, S. 2002 Towards a functional classification of the freshwater phytoplankton. *Journal of Plankton Research* **24**, 417–428.

Shannon, C. E. & Weaver, W. 1949 *The Mathematical Theory of Communication*. University of Illinois Press, Urbana, IL, USA, pp. 22–32.

Shi, H. 2004 *Names of Algae in Latin and Chinese*. China Ocean Press, Beijing, China.

Suikkanen, S., Laamanen, M. & Huttunen, M. 2007 Long-term changes in summer phytoplankton communities of the open northern Baltic Sea. *Estuarine, Coastal and Shelf Science* **71**, 580–592.

Tang, X., Wu, M., Yang, W., Yin, W., Jin, F., Ye, M., Currie, N. & Scholz, M. 2011 Ecological strategy for eutrophication control. *Water, Air, & Soil Pollution* **223**, 723–737.

Tang, C., Yi, Y., Yang, Z. & Sun, J. 2016 Risk forecasting of pollution accidents based on an integrated Bayesian Network and water quality model for the South to North Water Transfer Project. *Ecological Engineering* **96**, 109–116.

Wan, Y., Zhang, L., Fu, X. & Jin, H. 2020 Assessment of eutrophication degree and its prevention in Danjiangkou Reservoir. *Journal of Beijing Normal University (Natural Science)* **56**, 275–281.

Wang, Q., Cao, J. & Liu, Y. 2008 *A Collection of Pictures of Algae in Shanghai Jiuduansha Wetland National Nature Reserve and Its Surrounding Waters*. Science Press, Beijing, China.

Wang, X., Zhang, M. & Yin, J. 2018 Composition and influential factors of phytoplankton function groups in Lake Chaohu. *Journal of Lake Sciences* **30**, 431–440.

Wei, F., Qi, W. & Sun, Z. 2002 *Monitor and Analysis Method for Water and Wastewater*, 4th edn. China Environmental Science Press, Beijing, China.

Yuan, Y., Jiang, M., Liu, X., Yu, H., Otte, M. L., Ma, C. & Her, Y. G. 2018 Environmental variables influencing phytoplankton communities in hydrologically connected aquatic habitats in the Lake Xingkai basin. *Ecological Indicators* **91**, 1–12.

Zaghloul, F. A.-R., Khairy, H. M. & Hussein, N. R. 2020 Assessment of phytoplankton community structure and water quality in the Eastern Harbor of Alexandria, Egypt. *The Egyptian Journal of Aquatic Research* **46**, 145–151.

Zhou, F. & Chen, J. 2004 *Pictures of Freshwater Micro-Organisms*. Chemical Industry Press, Beijing, China.

Zhu, J., Lei, X., Quan, J. & Yue, X. 2019 Algae growth distribution and key prevention and control positions for the Middle Route of the South-to-North water diversion project. *Water* **11**, 1851.

First received 11 June 2020; accepted in revised form 23 September 2020. Available online 9 October 2020

Water quality index and spatio-temporal perspective of a large Brazilian water reservoir

Karla Lorrane de Oliveira, Ramatisa Ladeia Ramos, Sílvia Corrêa Oliveira and Cristiano Christofaro

ABSTRACT

The water spatio-temporal variability of the Irapé Hydroelectric Power Plant reservoir and its main tributaries was evaluated by analysing the temporal trend of the main parameters and applying the Canadian Council of Ministers of the Environment Water Quality Index (CCME WQI), using data from 2008 to 2018. This reservoir is in Minas Gerais, Brazil, and covers an area of approximately 143 km^2 across seven municipalities. The dissolved iron (DFe) presented the highest percentage of standards violations (31.7% to 80.5%), most frequently in the reservoir tributaries. The Mann–Kendall tests indicated that the monitoring stations showed an increasing trend of 78.5% $N\text{–}NH_4^+$ and 64.1% DFe. During the evaluated period, the reservoir waters were classified as excellent (1.2%), good (61.3%), acceptable (29.5%), and poor (8.0%) according to the WQI for the proposed use. The poorest quality classes were more frequent in the tributaries, especially in the year 2009. The WQI seasonal assessment indicated a worsening during the rainy season in 57% of the stations, as a result of external material transport to the water bodies. The CCME WQI, in conjunction with temporal statistical analysis, contributed to the interpretation of the monitoring data, generating important information for reservoir water quality management.

Key words | environmental statistics, hydroelectric reservoir, spatio-temporal perspective, water quality index

Karla Lorrane de Oliveira
Ramatisa Ladeia Ramos (corresponding author)
Sílvia Corrêa Oliveira
Department of Sanitary and Environmental Engineering,
Federal University of Minas Gerais,
P.O. Box 1294, ZIP 30.270-901, Belo Horizonte, MG,
Brazil
E-mail: *lramos.eng@gmail.com*

Cristiano Christofaro
Department of Forestry Engineering,
Federal University of Jequitinhonha and Mucuri Valleys,
P.O. Box 5000, ZIP 39.100-000, Diamantina, MG,
Brazil

HIGHLIGHTS

- The large Brazilian reservoir and tributaries water was studied by statistical techniques in conjunction with a WQI.
- Mann-Kendall tests indicated the stations with an increasing trend in the parameters analyzed.
- The seasonal WQI indicated a worsening in the stations in the rainy seasons.
- CCME WQI and temporal statistical analysis generated information that can be used in reservoir management.

INTRODUCTION

Quality assessment is essential for the proper use of water in different human activities and is affected by natural and anthropogenic factors as well as by hydrological dynamics. Lentic ecosystems, such as reservoirs, have characteristics that affect the spatio-temporal scale, presenting a different dynamic to lotic ecosystems. Thus, programmes for monitoring physical, chemical, and biological parameters of water quality in these environments are indispensable for a

doi: 10.2166/ws.2020.374

better understanding and evaluation of water body conditions. However, these programmes generate a complex dataset, requiring specific tools for proper interpretation and evaluation of the temporal and spatial variability in the water quality parameters.

Several statistical techniques can be used to understand the temporal dynamics in water reservoirs, such as tests of seasonality and temporal trends of quality parameters (Penev *et al.* 2014). These analyses allow us to determine, statistically, if the values of a random variable are decreasing or increasing over a certain period. We can also determine the seasonality of these variables (Helsel & Hirsch 2002; Yenilmez *et al.* 2011).

The complexity of interpreting many quality parameters can be reduced by applying quality indices that enable simultaneous evaluation of several parameters as well as natural and anthropic influences on the aquatic ecosystem's environmental dynamics. The water quality index (WQI), developed by the Canadian Council of Ministers of the Environment (CCME), aims to assess the distance between the current water quality and the goal established by the water resource framework (CCME 2001). Some studies have demonstrated the applicability and contribution of WQI to water quality diagnosis (Rosemond *et al.* 2008; Tyagi *et al.* 2013). These benefits can be extended with their use in conjunction with other statistical analyses.

The Irapé Hydroelectric Power Plant (HPP), located in the state of Minas Gerais, Brazil, is in a semi-arid region at risk of desertification (Tomasella *et al.* 2018). It has a reservoir of approximately 143 km^2, whose waters are used by the population of seven municipalities. Population growth and climate change have increased concerns about the water quality in this reservoir. Thus, understanding the water quality spatio-temporal dynamics of the Irapé HPP Reservoir and its tributaries is highly relevant, and may support the actions of decision makers (Helsel & Hirsch 2002; Penev *et al.* 2014) while contributing to a more efficient use of its waters by the local population.

In this study, the spatio-temporal variability of surface water in the Irapé HPP Reservoir and its main tributaries was evaluated using statistical techniques, comparison with environmental standards, and application of the CCME WQI. The association between statistical analysis and quality indices allows the identification of the points under greatest anthropogenic pressure in the reservoir and its surroundings, evaluates natural influences, and helps understand the temporal dynamics of pollutants in the reservoir and its tributaries. Notably, the results of the study can be applied to other reservoirs and can aid in multiple water use analysis.

MATERIALS AND METHODS

Study area

The Irapé HPP (Presidente Juscelino Kubitscheck Hydroelectric Plant), located at 16°44′15″ S and 42°34′30″ W, was inaugurated in 2006 and has a 142.95 km^2 reservoir, covering seven municipalities. It has maximum total and useful volumes of 5,954.88 hm^3 and 3,689 hm^3, respectively, and an installed capacity of 399 MW. The operational water levels vary between 470.8 m and 510 m. The reservoir is in the Alto Jequitinhonha watershed (JQ1 Water Resource Management Units – WRMU), which has an area of 19,855 km^2, covering a total of 26 municipalities with a population of approximately 120,965 inhabitants. The Alto Jequitinhonha region predominantly produces forest products, specifically from *Eucalyptus* sp., agriculture, and livestock (Silva & Miranda 2015).

Water quality monitoring data

The secondary data used in the present study were obtained from the water quality monitoring carried out between 2008 and 2018 by the Companhia Energética de Minas Gerais (Cemig) at 14 sampling stations in the Irapé HPP Reservoir and its tributaries. The geographical location and description of the monitoring stations are shown in Figure 1 and Table 1, respectively.

The dataset includes 14 water quality parameters (Table 2) monitored quarterly. The samples were collected and analysed according to Cemig's Manual of Sampling Procedures and Water Analysis Methodologies (2009).

Water quality parameter limits recommended by legislation

The violation percentages for each database parameter that has a limit recommended in Brazilian legislation by CONAMA, Brazil's National Environmental Council (Brazil 2005), were

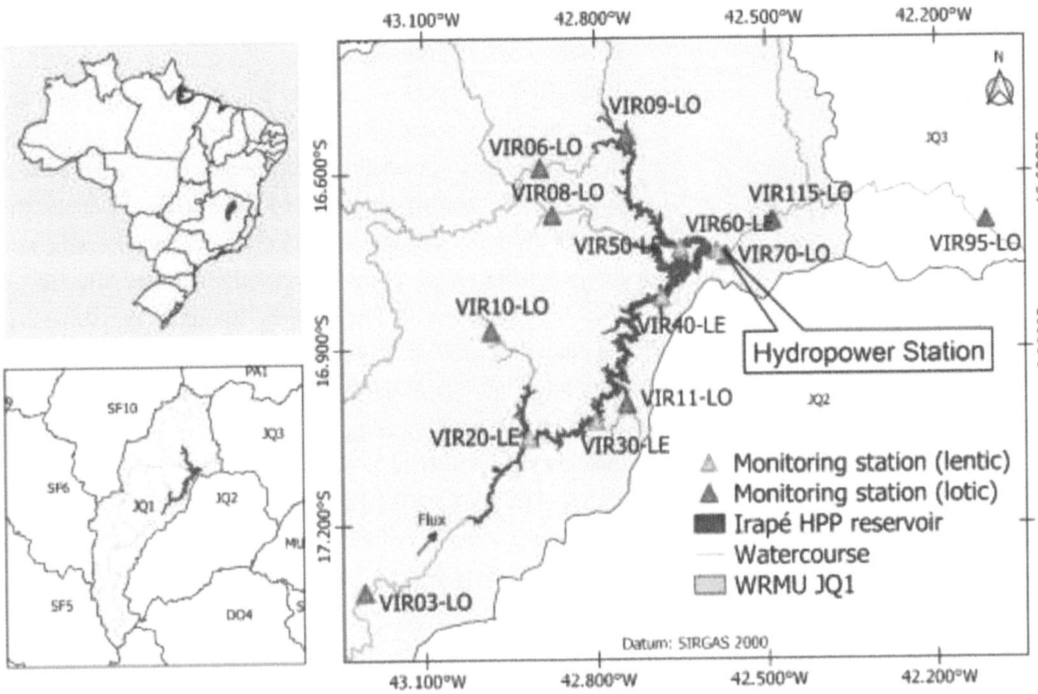

Figure 1 | Geographic location of the Irapé HPP Reservoir, its tributaries, and the water quality monitoring stations.

Table 1 | Description of the Irapé HPP Reservoir and its tributaries' monitoring stations

Station	Description	Watercourse	Município	Physical condition
VIR03-LO	Reservoir upstream – close to the Terra Branca ferry	Jequitinhonha River	Bocaiuva	Lotic
VIR06-LO	Reservoir upstream – below the bridge that connects Grão Mogol to Cristália	Itacambiruçu River	Grão Mogol	Lotic
VIR08-LO	Reservoir upstream – on the bridge that connects Grão Mogol to Cristália	Soberbo River	Cristália	Lotic
VIR09-LO	Reservoir upstream – on the bridge that connects Grão Mogol to Irapé HPP	Ventania River	Grão Mogol	Lotic
VIR10-LO	Reservoir upstream	Noruega River	Botumirim	Lotic
VIR11-LO	Reservoir upstream	Corrente River	Leme do Prado	Lotic
VIR70-LO	Downstream of the powerhouse – 500 m from the escape channel	Jequitinhonha River	Grão Mogol	Lotic
VIR95-LO	Downstream of the powerhouse – in front of Coronel Paulo Fernandes school	Jequitinhonha River	Coronel Murta	Lotic
VIR115-LO	Downstream of the powerhouse	Jequitinhonha River	Virgem da Lapa	Lotic
VIR20-LE	Reservoir	Jequitinhonha River	Turmalina	Lentic
VIR30-LE	Reservoir	Jequitinhonha River	Leme do Prado	Lentic
VIR40-LE	Reservoir	Jequitinhonha River	Grão Mogol	Lentic
VIR50-LE	Reservoir	Itacambiruçu River	Cristália	Lentic
VIR60-LE	Reservoir, 500 m from the dam	Jequitinhonha River	Grão Mogol	Lentic

Table 2 | Surface water quality and standards set by CONAMA

Parameter	Abbreviation	Unit	Standard
Total alkalinity	TAlc	$mg{\cdot}L^{-1}$ $CaCO_3$	*
Biochemical oxygen demand	BOD	$mg{\cdot}L^{-1}$ O_2	5
Dissolved oxygen	DO	$mg{\cdot}L^{-1}$ O_2	>5
Dissolved iron	DFe	$mg{\cdot}L^{-1}$ Fe	0.3
Electrical conductivity	EC	$\mu S.cm^{-1}$	*
Total ammoniacal nitrogen	N-NH_4^+ pH $\leq$ 7.5	$mg{\cdot}L^{-1}$ N	3.7
	N-NH_4^+ 7.5 < pH < 8.0	$mg{\cdot}L^{-1}$ N	2
	N-NH_4^+ 8.0 < pH < 8.5	$mg{\cdot}L^{-1}$ N	1
	N-NH_4^+ pH $\geq$ 8.5	$mg{\cdot}L^{-1}$ N	0.5
Nitrate	N-NO_3^-	$mg{\cdot}L^{-1}$ N	10.0
pH in loco	pH	–	6 to 9
Water temperature	Temperature	°C	*
Thermotolerant coliforms	Therm. coli.	org.100 Ml^{-1}	1,000
Total dissolved solids	TDS	$mg{\cdot}L^{-1}$	500
Total phosphorus	TP (lotic)	$mg{\cdot}L^{-1}$ P	0.1
	TP (lentic)	$mg{\cdot}L^{-1}$ P	0.03
Sulfate	SO_4^{-2} T	$mg{\cdot}L^{-1}$ P	250
Turbidity	Turb.	NTU	100

*No limit recommended by CONAMA.

calculated according to the watercourse classification for each station (Class 2). The limits are listed in Table 2.

Trend analysis

The temporal trend analysis of the parameters was performed by station, using the Mann–Kendall test (MK) or Mann–Kendall Seasonal test (MKS), commonly used in temporal analysis of environmental data owing to its simplicity and robustness (Yenilmez *et al.* 2011). One assumption for the tests that produced reliable results was the lack of autocorrelation in the analysed data. In this study, the autocorrelation was verified through the autocorrelation function (ACF), which measures the degree of variable correlation at a given moment, with itself and at a later point in time.

The choice between MK and MKS was based on the presence or absence of seasonality within the data measured at different periods of the year, as this factor is a potential source of variation in the water quality data series (Helsel & Hirsch 2002). Seasonality was analysed using the Kruskal–Wallis (KW) non-parametric statistical test, applied to the quarterly data for all parameters in each season. When $p < 0.05$ in the KW test, the seasonality influence was considered to exist and the MKS test was applied. In cases where $p > 0.05$, the seasonality influence was considered to be non-existent and the parameter temporal trend was verified by the MK test.

The time trend was verified by Kendall's tau values (τ) calculated in MKS or MK. For $p < 0.05$, the trend was considered to exist, and the Kendall tau value (τ) determined whether this trend was upward (positive τ) or downward (negative τ). Statistical tests were performed using the programs XLSTAT® 2014.1.01 and/or Statistica® 8.0, at a significance level (α) of 5%.

CCME WQI

The Brazilian legal limits set by CONAMA (Brazil 2005) were used in the CCME WQI calculation. The index is a combination of three factors that represent non-compliance with the proposed quality criteria to produce a single value (between 0 and 100) that describes water quality (CCME 2001).

F1 (scope) represents the percentage of variables that did not meet the objectives at least once during the time period under consideration ('failed variables'), relative to the total number of variables measured (Equation (1)):

$$F_1 = \left(\frac{Number\ of\ failed\ variables}{Total\ number\ of\ variables}\right) \times 100 \tag{1}$$

F2 (frequency) represents the percentage of individual tests that did not meet the objectives ('failed tests') (Equation (2)):

$$F_2 = \left(\frac{Number\ of\ failed\ tests}{Total\ number\ of\ tests}\right) \times 100 \tag{2}$$

F3 (amplitude) represents the amount by which failed test values did not meet their objectives. It is calculated in three steps.

(1) The number of times by which an individual concentration is greater than (or less than, when the objective

is a minimum) that of the objective is termed an 'excursion' and is expressed as follows.

When the test value did not exceed the objective (Equation (3)):

$$\Delta v = \left(\frac{failed\ test\ value}{objective}\right) - 1 \quad (3)$$

For cases in which the test value did not fall below the objective (Equation (4)):

$$\Delta v = \left(\frac{objective}{failed\ test\ value}\right) - 1 \quad (4)$$

(2) The collective amount by which individual tests are out of compliance is calculated by summing the excursions of individual tests from their objectives and dividing it by the total number of tests (all those meeting objectives and not meeting objectives). This variable, referred to as the normalised sum of excursions (NSE) is calculated using Equation (5):

$$NSE = \frac{\sum_{i=1}^{n} excursion}{total\ number\ of\ tests} \quad (5)$$

(3) F3 is then calculated using an asymptotic function that scales the normalised sum of the excursions from objectives (nse) to yield a range between 0 and 100 (Equation (6)):

$$F_3 = \frac{NSE}{(0,01 \times NSE) + 0,01} \quad (6)$$

Finally, the WQI can be calculated using Equation (7):

$$WQI = 100 - \left(\frac{\sqrt{F_1^2 + F_2^2 + F_3^2}}{1{,}732}\right) \quad (7)$$

The index ranges from 0 to 100 were divided into five categories by the CCME: (i) excellent (95–100), (ii) good (80–94), (iii) reasonable (65–79), (iv) marginal (45–64), and (v) terrible (0–44). The WQI methodology proposed by CCME does not define the parameters but recommends a minimum of eight and a maximum of 20 parameters to be used in the calculation (CCME 2012). In this study, all monitored parameters that defined legal standards were considered (Table 2). The WQI was applied to each monitoring station and to each year. In the second step, the index was calculated by season, dry (April to September) and rainy (October to March), to analyse the influence of seasonality on the water quality.

RESULTS AND DISCUSSION

Legislation standards violations

Box-plot graphs with the standards violations of each parameter for all seasons are shown in Figure 2.

The legal limit violations registered for the pH occurred below the minimum allowed, highlighting the value of 2.99 recorded in 2017 for the station VIR10-LO. Acidic pH may be associated with the presence of red oxisols in the watershed, which in general are strongly acidic soils. In addition, the contact of aerated water from the reservoir with the dam, mainly a homogeneous pack of quartz-mica-shale containing sulphides disseminated in the rock matrix (Duarte *et al.* 2009), leads to the oxidation of sulphide minerals, resulting in low pH solutions that cause regional water acidification. Low pH values can affect growth or cause death in ichthyofauna. Wide pH variability can affect more sensitive organisms, even if the legal limits are not exceeded. In addition, low pH values can solubilise other metals, increasing environmental toxicity (Rodrigues 2002).

The high frequencies of dissolved Fe violations at all monitoring stations (31.7% to 80.5%) may be related to the local geochemical and pedological characteristics, such as the presence of red oxisols, which have a high Fe_2O_3 content. An association between high iron content and soil characteristics was observed in the Nisa River (Czech Republic and Germany) (Kändler *et al.* 2017). The reservoir waters' acidic pH can also contribute by reducing Fe^{+3} (insoluble) in Fe^{+2} (dissolved).

The high percentage of DO violation at the VIR70-LO station may be related to its location downstream from the powerhouse. The water intake of the HPP is 40 m deep,

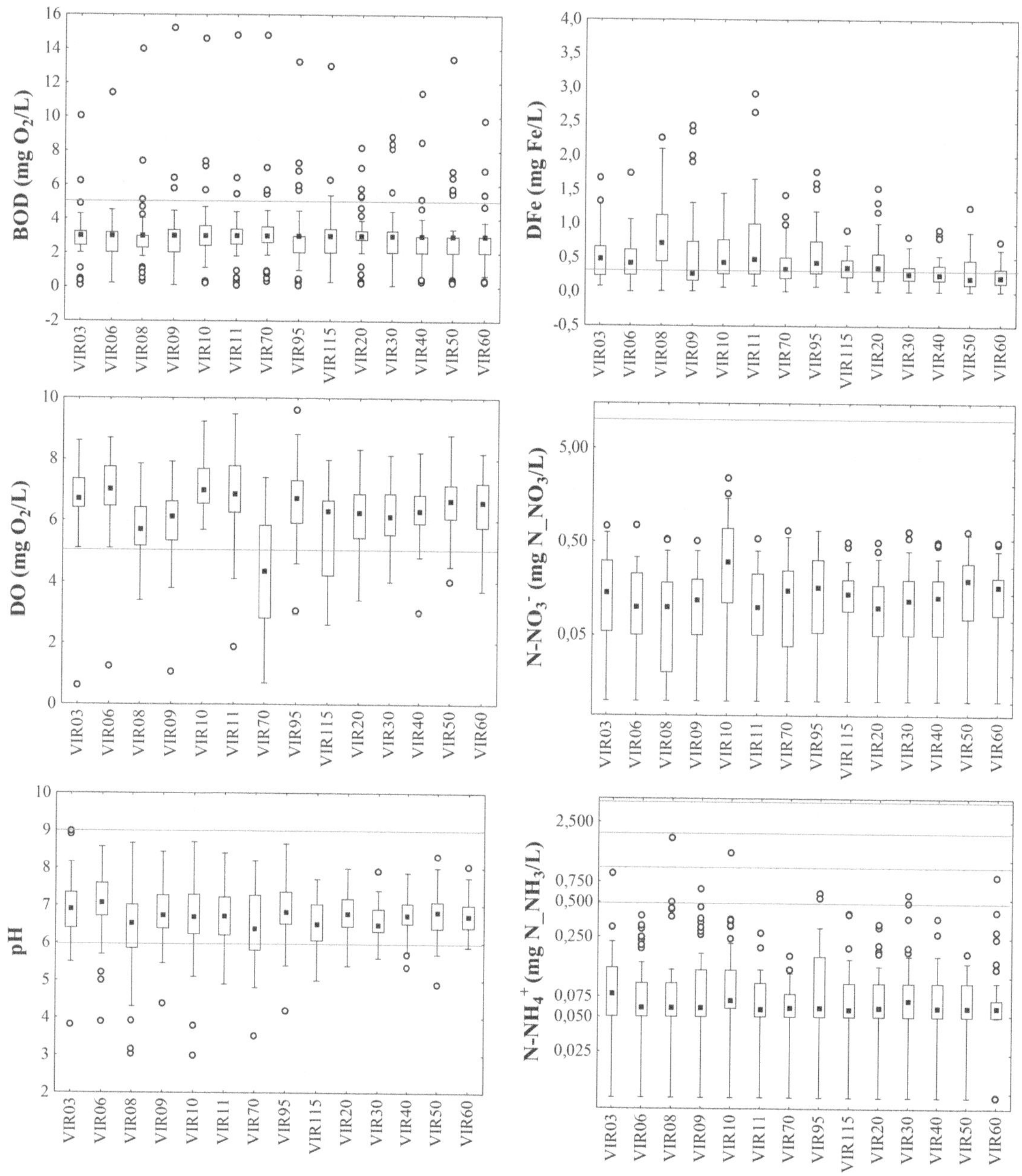

Figure 2 | Variability of water quality parameters and legal standards violations of the monitoring stations of Irapé HPP Reservoir and its main tributaries. *Note*: The gray horizontal line represents the legal standard or standard range of each parameter.

and thus the water returned from the machines can be characterised as bottom water, which in general has lower DO values than those of surface water. The low DO values should be monitored as this is an extremely important parameter for aerobic aquatic organism respiration and has been used in the calculation of various WQIs since the 1960s (Tyagi *et al.* 2013). In addition, the decrease in DO concentrations tends to influence chemical changes in the nutrient forms and metals, in addition to other water quality parameters, which may become available in the environment and/or have their values changed (Ashby 2009).

The parameters N–NO_3, N–NH_4^+, and SO_4^{-2} T did not exceed the legal limits, while the TDS showed a violation in a single sampling campaign at station VIR20-LE (2.4%).

However, the pH, BOD, and TP limits were violated at least once in every season and in all years of the historical series, with percentages varying between 4.8% and 34.9%; 2.4% and 14.3%; and 12.2% and 26.8%, respectively. BOD and TP are important indicators of organic pollution in water bodies. High BOD loads mainly originate from anthropogenic sources, including domestic and animal waste, industrial emissions, and sewage releases (Vigiak *et al.* 2019). These high values indicate a great demand for oxygen that is necessary to break down the organic matter present in the water body through bacterial respiratory activity, consequently decreasing the concentration of available DO. The bioavailable forms of phosphorus, together with inorganic nitrogen, play an important role in aquatic ecology. Excess phosphorus can lead to eutrophication of water resources, which can cause ecological and toxicological effects that are directly or indirectly related to primary producer proliferation (Wang *et al.* 2009; Moal *et al.* 2019).

For TP, 78% of the values above the limit occurred during the rainy season, indicating a greater transport of allochthonous material to the reservoir during this period by surface runoff. Yenilmez *et al.* (2011) and Li *et al.* (2019) found a statistically positive relationship between TP and precipitation when analysing the water quality of Lake Eymir in Turkey and the Three Gorges Reservoir in China, respectively, concluding that the parameter is introduced into the lake by runoff. In Irapé, the effects of agricultural or silvicultural practices in the tributary area, the main anthropic uses observed in the watershed (Silva & Miranda 2015), could be responsible for the TP contribution. The turbidity behaviour, with higher percentages of violation in the tributaries, although without the occurrence of values above the legal limits, reinforces this hypothesis. The reservoirs' low turbidity can be explained by the sedimentation effect, as also observed by Li *et al.* (2019) in the Three Gorges Reservoir region, China, and in the Nova Ponte Reservoir, Brazil by Christofaro *et al.* (2017).

Thermotolerant coliforms have a higher percentage of standards violations in the tributaries in relation to the reservoir. This parameter has a strong connection with point sources of untreated sanitary sewage, indicating that the tributaries are affected by this type of pollution. The highest percentage of violation of thermotolerant coliforms occurred at the VIR11-LO station (30%), located on the right arm of the main river, indicating that the flow rates and conditions for regeneration in the region reduce the impact of domestic sewage release (even if untreated).

Trend analysis of the reservoir water quality parameters and main tributaries

The autocorrelation coefficient did not reach a significant value for most parameters, except for isolated occurrences of autocorrelation for SO_4^{-2} T (Figures S1 to S14). Thus, the Mann–Kendall tests (MK or MKS) were used for all parameters at all stations.

The temperature showed a substantial seasonal variation, with values significantly higher ($p < 0.05$) in the rainy season in 11 stations. For the other parameters, seasonal differences, when detected, indicated higher levels during the rainy season (Table 3).

The N–NH_4^+ and DFe were the parameters that showed the highest time trend occurrences, detected in 78.6% and 64.3% of the monitoring stations, respectively. The tendency to increase N–NH_4^+ may be associated with a greater contribution from agricultural sources in the basin, which may result in greater toxicity to aquatic organisms as well as a reduction in DO concentration in the water. However, the concentrations of N–NH_4^+ did not exceed the legal standards. On the other hand, the DFe has a significant impact and a tendency to increase the reservoir water quality and its tributaries, which can be associated with an increase in soil exposure in the watershed over the study period.

Turbidity showed an upward trend at eight stations (57.1%), with five inside the reservoir. Although there are still no representative standards violations, the results indicate the need to adopt measures to control turbidity, especially in relation to tributaries and the reservoir surroundings.

Only TDS and thermotolerant coliforms showed significant reduction trends. All stations located in the reservoir presented a temporal tendency to reduce thermotolerant coliforms, two other stations are on the main river (Jequitinhonha River), immediately downstream of the water mirror (VIR70-LO and VIR115-LO). These results indicate that the point source effects of sanitary sewage on reservoir waters decreased over the study period.

Table 3 | Influence of seasonality and temporal trends of the analysed parameters in each monitoring station of the Irapé HPP Reservoir and its main tributaries

Parameters

Stations	TAlc*	Therm. coli.	EC*	BOD	DFe	TP	$N\text{-}NO_3^-$	$N\text{-}NH_4^+$	DO	pH	TDS	SO_4^{-2} T	Temp*	Turb.
VIR03-LO		W (17.1%)		(4.9%)	(63.4%)	W (22.0%)	W (0.0%)	(0.0%)	D (2.4%)	(11.9%)	↓ (0.0%)	(0.0%)	W	↑ (17.5%)
VIR06-LO	D	W (22.5%)		↑ (2.4%)	W (68.3%)	W (20.0%)	W (0.0%)	↑ (0.0%)	(2.5%)	(9.8%)	(0.0%)	(0.0%)		W (18.0%)
VIR08-LO		(12.5%)		↑ (9.8%)	↑ (80.5%)	(15.4%)	↓ (0.0%)	(0.0%)	(22.5%)	(31.7%)	↓ (0.0%)	(0.0%)	W	W (7.9%)
VIR09-LO		(24.3%)		(9.8%)	↑ (43.9%)	(15.8%)	↑ (0.0%)	↑ (0.0%)	(22.5%)	(12.2%)	(0.0%)	(0.0%)	W	↑ (5.4%)
VIR10-LO		W (17.5%)	↑	(9.8%)	↑ (68.3%)	(22.5%)	D (0.0%)	(0.0%)	(0.0%)	↓ (19.5%)	(0.0%)	(0.0%)	↑	(5.6%)
VIR11-LO		W (30.0%)	↑	(9.5%)	↑ (72.5%)	(20.0%)	(0.0%)	↑ (0.0%)	(7.3%)	(16.7%)	(0.0%)	(0.0%)		W (5.3%)
VIR70-LO		W ↓ (13.2%)		(10.0%)	↑ (58.1%)	(14.3%)	(0.0%)	↑ (0.0%)	↑ (66.7%)	↑ (34.9%)	(0.0%)	(0.0%)	D	↑ (2.6%)
VIR95-LO		W (24.4%)		↑ (12.2%)	↑ (63.4%)	(24.4%)	W (0.0%)	↑ (0.0%)	(5.0%)	(12.2%)	(0.0%)	(0.0%)	W	W (26.3%)
VIR115-LO		↓ (15.8%)		(8.6%)	(60.5%)	(13.2%)	(0.0%)	↑ (0.0%)	↑ (27.0%)	(18.4%)	↓ (0.0%)	(0.0%)	W	(5.9%)
VIR20-LE		↓ (2.8%)		(12.5%)	↑ (56.4%)	(26.8%)	(0.0%)	↑ (0.0%)	(5.0%)	D (14.6%)	(2.4%)	(0.0%)	W	↑ (0.0%)
VIR30-LE	W	↓ (2.7%)		(10.0%)	↑ (46.3%)	(17.1%)	(0.0%)	↑ (0.0%)	(12.5%)	(10.0%)	↓ (0.0%)	(0.0%)	W	↑ (0.0%)
VIR40-LE	W	↓ (2.9%)		↑ (7.1%)	(39.0%)	↑ (12.2%)	(0.0%)	↑ (0.0%)	(4.9%)	(7.1%)	↓ (0.0%)	(0.0%)	W	↑ (0.0%)
VIR50-LE		↓ (5.7%)		(14.3%)	↑ (38.1%)	(23.1%)	(0.0%)	↑ (0.0%)	(4.9%)	(14.3%)	↓ (0.0%)	(0.0%)	W	↑ (0.0%)
VIR60-LE		↓ (0.0%)		(7.1%)	(31.7%)	(17.1%)	(0.0%)	↑ (0.0%)	(2.4%)	(4.8%)	(0.0%)	(0.0%)	W	↑ (0.0%)

(↑) = upward trend; (↓) = downward trend; () = no trend; (W) = influence of seasonality with higher values observed in the rainy season; (D) = influence of seasonality with higher values observed in the dry season; (*) = no standard limit in the legislation.

CCME WQI application

Table 4 shows the CCME WQI values obtained in the seasons for the years in which there were quarterly samplings.

WQI values ranged between 54.68 and 100.00. Of the 88 indices calculated, one point, in 2014 (VIR06-LO), was classified as excellent, 61.3% were classified as good, indicating that they rarely differed from the legal limits, and 29.5% were acceptable, which sometimes went over the legislation limits. Only 8.0% fell into the bad range (they often violated legal limits), with no record of the very poor class. The monitoring points located inside the reservoir were generally classified as good, except for in 2009.

The better water quality at the reservoir's monitoring points indicated that the water body reduced the effects of pollution from tributaries, as observed by Xing *et al.* (2015), when comparing the water quality of the Danjiangkou Reservoir in China with the rivers that supply it. This reduction can be caused by changes in river dynamics, such as a decrease in water flow velocity and the consequent deposition of pollutants by sedimentation, a phenomenon common to lentic environments (Wang *et al.* 2009; Li *et al.* 2019).

Gao *et al.* (2016) also found similar results for the Three Gorges Reservoir, the largest hydroelectric project in the world, in which the decrease in concentrations of heavy metals from upstream to downstream was associated with reservoir self-purification, which was stable and acceptable from 2008 to 2013. Other studies have also shown that the CCME WQI is a valuable means of monitoring, communicating, and understanding surface water quality (Hurley *et al.* 2012; Ahmed *et al.* 2020).

Table 5 presents the WQI seasonality, considering the dry and rainy seasons. There is a deterioration of the index values in the rainy season (October to March), indicating a greater supply of nutrients with rain during this period, suggesting a predominance of diffuse pollution sources (Barbosa *et al.* 2019). In addition, the amplitude of reservoir seasonal variation was less than that observed for the tributaries.

The existing impacts on the water quality of the Irapé Reservoir have natural and anthropic origins and they can be intensified during rainy seasons. However, even though some problems have a natural origin, they can be aggravated by human actions, such as inadequate soil management in agricultural practices in the region. One strategy that can be adopted to minimise this problem is investment in reforestation and environmental restoration, where the vegetation acts as a filter and barrier for rainwater surface runoff,

Table 4 | CCME WQI by season and year, applied to the Irapé HPP monitoring database and main tributaries

Station	Year										
	2008	2009	2010	2011	2012	2013	2014	2015	2016	2017	2018
VIR03-LO	-	76.22	-	82.11	85.72	-	93.90	94.03	86.07	-	69.62
VIR06-LO	-	76.01	-	77.83	87.15	-	100.00	-	81.25	-	81.65
VIR08-LO	-	63.94	-	59.74	84.19	-	94.18	-	70.83	-	73.08
VIR09-LO	-	72.19	-	63.03	89.14	-	88.62	-	72.34	-	86.29
VIR10-LO	-	70.21	-	77.34	76.75	-	83.69	-	89.59	-	82.86
VIR11-LO	-	68.75	-	82.14	73.82	-	82.24	-	74.05	-	77.69
VIR70-LO	-	62.69	-	81.25	70.59	-	72.02	-	80.44	-	-
VIR95-LO	-	72.07	-	88.69	68.16	-	89.17	69.21	62.81	-	82.39
VIR115-LO	-	54.68	-	83.32	82.54	-	-	-	75.27	-	-
VIR20-LE	-	60.04	-	87.46	80.65	-	83.73	81.97	79.77	-	80.24
VIR30-LE	-	70.06	-	86.45	82.36	-	83.68	94.58	92.72	-	86.60
VIR40-LE	-	71.76	-	93.74	91.35	-	94.58	91.51	88.22	-	88.03
VIR50-LE	-	71.51	-	87.17	86.62	-	83.69	87.90	90.64	-	88.21
VIR60-LE	-	71.94	-	92.99	89.70	-	88.70	94.50	91.11	-	94.10

Excellent   Good   Acceptable   Bad   Very bad   (-) no quarterly sampling

Table 5 | WQI by season, applied to the Irapé HPP monitoring database and main tributaries

	Monitoring stations						
Season	VIR03-LO	VIR06-LO	VIR08-LO	VIR09-LO	VIR10-LO	VIR11-LO	VIR70-LO
Rainy	66.11	67.66	64.27	58.38	63.19	65.27	58.99
Dry	93.75	94.05	70.83	76.96	78.13	78.08	71.43

	Monitoring stations						
Season	VIR95-LO	VIR115-LO	VIR20-LE	VIR30-LE	VIR40-LE	VIR50-LE	VIR60-LE
Rainy	59.52	66.11	72.3	73.03	89.25	77.17	83.95
Dry	66.83	69.98	76.06	73.96	81.67	88.69	81.51

Good Acceptable Bad

preventing erosion caused by the direct impact of raindrops on the soil.

In this sense, Li *et al.* (2019) note that the Chinese government's investments in reforestation and environmental restoration has positive impacts on reservoir water quality. In addition, it is important to emphasise that actions for the management of this and other reservoirs include the awareness and environmental education of the region's population, including promoting the dissemination of and access to information related to water quality. It should also be noted that management must occur in an integrated manner, based on an understanding of the structure and functioning of the reservoir as an ecosystem, through the use of surveys and soil occupation in its surroundings, and with the help of the local communities.

CONCLUSION

High Fe concentrations and a large percentage of stations with an increasing trend indicate the increase of soil exposure in the watershed. Despite not exceeding legal standards, $N\text{–}NH_4^+$ showed an upward trend in 64% of the monitoring stations, which may be related to an increase in the contribution of agricultural runoff. The agricultural use of the soil can also be related to the TP contribution to the reservoir and tributaries and to the turbidity values that violated the limit in the tributaries. The occurrence of low pH values must be monitored, and they may be related to the presence of iron sulphides in the region. The thermotolerant coliforms showed a downward trend, mainly in the reservoir monitoring stations, indicating a reduction in the effect of point sources of sanitary sewage discharge.

The reservoir water quality and its tributaries were considered adequate during the study period, with approximately 61% of the 88 WQI qualifying as good, 29.5% as acceptable, and 8.0% as bad. The reservoir points are of better quality than the tributaries, and the WQI assessment showed a deterioration in the rainy season, reinforcing the influence of diffuse sources on water quality. The application of the CCME WQI together with temporal statistical analysis showed a potential contribution to the interpretation of environmental monitoring data, generating important information for reservoir water quality management. Therefore, the results can support control and management measures in specific stretches of watersheds, with a focus on the most relevant polluting sources.

ACKNOWLEDGEMENTS

The authors would like to thank the Energy Company of Minas Gerais State (Cemig), the Foundation for Research Support of the State of Minas Gerais (FAPEMIG), the Coordination for the Improvement of Higher Education Personnel (CAPES), and the National Council of Technological and Scientific Development (CNPq) for their support and cooperation during the research.

DATA AVAILABILITY STATEMENT

All relevant data are included in the paper or its Supplementary Information.

REFERENCES

Ahmed, S., Shadabkhurshid, K., Madan, R., Abu Amarah, B. A. & Naushad, M. 2020 Water quality assessment of shallow aquifer based on Canadian Council of Ministers of the Environment index and its impact on irrigation of Mathura District, Uttar Pradesh. *Journal of King Saud University* **32** (1), 1218–1225. https://doi.org/10.1016/j.jksus.2019.11.019.

Ashby, S. 2009 Impacts of hydrology and hydropower on water quality in reservoir tailwaters. *WIT Transactions on Ecology and the Environment* **124**, 55–56. doi:10.2495/RM090061.

Barbosa, A. G., Santos, K. M. S. & Michelan, D. C. G. S. 2019 Avaliação da sazonalidade da qualidade da água do açude da Macela em Itabaiana/SE. *Scientia cum Industria* **7** (3), 24–31. http://dx.doi.org/10.18226/23185279.v7iss3p24.

Brazil 2005 *Resolução N. 357, de 17 de Março de 2005.* National Environmental Council (CONAMA), Environmental Guidelines for Water Resources and Standards for the Release of Effluents (in Portuguese).

CCME – Canadian Council of Ministers of the Environment 2001 *Canadian Sediment Quality Guidelines for the Protection of Aquatic Life – Polichlorinated Dibenzo-P-Dioxins and Polychlorinated Dibenzofurans (PCDD/Fs).* CCME, Ottawa.

CCME – Canadian Council of Ministers of the Environment 2012 *Synthesis of Research and Application of the CCME Water Quality Index.* CCME, Winnipeg.

Cemig – Companhia Energética de Minas Gerais 2009 *Manual de Procedimentos de Coleta E Metodologias de Análise de Água: SISÁGUA.* Cemig, Belo Horizonte, p. 77.

Christofaro, C., Leão, M. M. D., Oliveira, S. C., Viana, D. T., Amorim, C. C. & Carvalho, M. D. 2017 Spatio-temporal variations of water quality in Nova Ponte Reservoir, Araguari River Basin, Brazil. *Water Science & Technology: Water Supply* **17** (6), 1507–1514. https://doi.org/10.2166/ws.2017.055.

Duarte, J., Filho, P. L. M., Calcina, A. M. & Wunder, E. 2009 Underground works of the Irapé power plant in Brazil in a sulfide-rich rock mass – special features and measures for concrete protection. In: *ITA World Tunnel Congress*, Budapest.

Gao, Q., Li, Y., Cheng, Q., Yu, M., Hu, B., Wang, Z. & Yu, Z. 2016 Analysis and assessment of the nutrients, biochemical indexes and heavy metals in the Three Gorges Reservoir, China, from 2008 to 2013. *Water Research* **92**, 262–274. https://doi.org/10.1016/j.watres.2015.12.055.

Helsel, D. R. & Hirsch, R. M. 2002 *Statistical Methods in Water Resources.* Elsevier Science Publishers B.V, Amsterdam, p. 525. https://doi.org/10.3133/twri04A3

Hurley, T., Sadiq, R. & Mazumder, A. 2012 Adaptation and evaluation of the Canadian Council of Ministers of the Environment Water Quality Index (CCME WQI) for use as an effective tool to characterize drinking source water quality. *Water Research* **46** (11), 3544–3552. https://doi.org/10.1016/j.watres.2012.03.061.

Kändler, M., Blechinger, K., Seidler, C., Pavlů, V., Šanda, M., Dostál, T., Krása, J., Vitvar, T. & Štich, M. 2017 Impact of land use on water quality in the upper Nisa catchment in the Czech Republic and in Germany. *Science of the Total Environment* **586**, 1316–1325. https://doi.org/10.1016/j.scitotenv.2016.10.221.

Li, Z., Ma, J., Guo, J., Paerl, H. W., Brookes, J. D., Xiao, Y., Fang, F., Ouyang, W. & Lu, L. 2019 Water quality trends in the Three Gorges Reservoir region before and after impoundment (1992–2016). *Ecohydrology & Hydrobiology* **19** (3), 317–327. https://doi.org/10.1016/j.ecohyd.2018.08.005.

Moal, M. L., Gascuel-Odoux, C., Ménesguen, A., Souchond, Y., Étrillard, C., Levain, A., Moatard, F., Pannard, A., Souchu, F., Lefebvre, A. & Pinaya, G. 2019 Eutrophication: a new wine in an old bottle? *Science of The Total Environment* **651**, 1–11.

Penev, S., Leonte, D., Lazarov, Z. & Mann, R. A. 2014 Applications of MIDAS regression in analysing trends in water quality. *Journal of Hydrology* **511**, 151–159.

Rodrigues, L. M. 2002 *Alterações Espaciais E Temporais de Características Limnológicas Resultantes da Transformação rio-Reservatório na área do Reservatório da Usina Hidrelétrica Dona Francisca, RS, Brasil.* Master's thesis, Universidade Federal do Rio Grande do Sul, Porto Alegre, p. 100.

Rosemond, S., Duro, D. & Dubé, M. 2008 Comparative analysis of regional water quality in Canada using the Water Quality Index. *Environmental Monitoring and Assessment* **156** (223). https://doi.org/10.1007/s10661-008-0480-6

Silva, T. V. & Miranda, G. H. R. 2015 *Análise das Variáveis da Geografia Física E Humana sob A Mesorregião do Vale do Jequitinhonha.* Uberlândia.

Tomasella, J., Vieira, R. M. S. P., Barbosa, A. A., Rodriguez, D. A., Santana, M. O. & Sestini, M. F. 2018 Desertification trends in the Northeast of Brazil over the period 2000–2016. *International Journal of Applied Earth Observation and Geoinformation* **73**, 197–206. https://doi.org/10.1016/j.jag.2018.06.012.

Tyagi, S., Sharma, B., Singh, P. & Dobhal, R. 2013 Water quality assessment in terms of water quality index. *American Journal of Water Resources* **1** (3), 34–38. doi:10.12691/ajwr-1-3-3.

Vigiak, O., Grizzetti, B., Udias-Moinelo, A., Zanni, M., Dorati, C., Bouraoui, F. & Alberto, P. 2019 Predicting biochemical oxygen demand in European freshwater bodies. *Science of The Total Environment* **666**, 1089–1105.

Wang, Y., Shen, Z., Niu, J. & Liu, R. 2009 Adsorption of phosphorus on sediments from the Three-Gorges Reservoir

(China) and the relation with sediment compositions. *Journal of Hazardous Materials* **162** (1), 92–98. https://doi.org/10.1016/j.jhazmat.2008.05.013.

Xing, X. K., Li, K. F., Finlayson, B. & Yin, W. 2015 Evaluation, prediction, and protection of water quality in Danjiangkou Reservoir, China. *Water Science and Engineering* **8** (1), 30–39. https://doi.org/10.1016/j.wse.2014.11.001.

Yenilmez, F., Keskin, F. & Aksoy, A. 2011 Water quality trend analysis in Eymir Lake, Ankara. *Physics and Chemistry of the Earth* **36**, 135–140. https://doi.org/10.1016/j.pce.2010.05.005.

First received 23 August 2020; accepted in revised form 9 December 2020. Available online 28 December 2020

Construction and application research of ecological evaluation platform based on Cesium

Yao Fu, Guize Luan, Jingzhi Cai, Yuchen Li and Fei Zhao

ABSTRACT

In this study, Sentinel 2A and Landsat series remote sensing images were used to determine the weights of vegetation coverage, humidity, dryness and heat by using the principal component analysis method. The remote sensing ecological index evaluation model was used to evaluate the ecology of the Fuxian Lake area from 2013 to 2018. Environmental quality in the study area is monitored and analyzed. By developing a unified organization and management module of spatio-temporal data in the collaborative production of multi-source heterogeneous data, the mapping relationship between data and parameters is established. The 3D-web data integration decision platform is constructed. Based on the Cesium engine real-time rendering and interactive visualization network 3D platform, real-time calculation and visualization of the ecological evaluation of the Fuxian Lake area between 2013 and 2018 is realized. The ecological environment of the study area continued to deteriorate from 2013 to 2015, improved slightly in 2016, and improved significantly in 2017–2018. The platform has the functions of searching, selecting the base map, measuring on the map, and positioning the coordinates, which can achieve the real-time ecological analysis function and show the change process of the ecological environment and natural resources.

Key words | data decision platform, ecological analysis function, map service, remote sensing ecological index

Yao Fu
Guize Luan
Jingzhi Cai
Yuchen Li
Fei Zhao (corresponding author)
College of Earth Sciences,
Yunnan University,
Kunming, Yunnan 650091,
China
E-mail: *vwobai@163.com*

HIGHLIGHTS

- Remote sensing data were used to determine the weights of *RSEI* by using the principal component analysis method.
- Environmental quality is monitored and analyzed.
- The 3D-web data integration decision platform is constructed.
- A real-time rendering and interactive visualization network 3D platform based on the Cesium engine is constructed.
- The platform has many functions.

INTRODUCTION

At present, the relationship between human activities and the ecological environment is getting closer and closer. Human activities have caused profound changes and influences on the ecological environment and ecosystem, and the resulting contradictions among population, resources and environment are becoming increasingly prominent. Therefore, the study of ecological security is of great significance for understanding the status of ecological security and for making relevant ecological environment decisions. At present, China's ecological environment deterioration tendency has

doi: 10.2166/ws.2020.253

not been effectively controlled in general. Although there has been partial improvement in recent years, the degree of governance is still far from recovering the serious consequences caused by ecological damage. With the proposal of national ecological civilization construction, the implementation of a sustainable development strategy of harmonious coexistence between humans and nature, and the rapid development of science and technology informatization levels in recent years, society is in urgent need of the introduction of new technologies and the application of new technologies in the improvement and protection of the ecological environment (Mitrović 2015). The formation and development of an ecosystem will be affected by multiple factors, and a single ecological factor cannot objectively and comprehensively reflect the changes of its ecological environment (Song & Xue 2016). Therefore, it is particularly important to analyze the comprehensive effects of multiple ecological factors and comprehensively evaluate the regional ecological environment. In recent years, a new remote sensing ecological index (*RSEI*) has been proposed. It is completely based on remote sensing information, which can directly represent regional ecological quality, intuitively reflect multiple indicators of the ecological environment, and realize rapid monitoring and evaluation of the regional ecological environment, with certain practical significance and high practical reference value (Lintao *et al.* 2014). At present, the application of *RSEI* is mainly limited to calculating green degree, dryness degree, humidity and heat indexes respectively through the information values of different bands of remote sensing images, and then synthesizing these indexes to finally reflect the changes of ecological environment in the study area from some angles. Some scholars have used a feature inversion technology and improved remote sensing ecological index to predict the ecological change caused by population growth in the Xiongan new area (Xu *et al.* 2018, 2019). Some scholars have made a comparative analysis of the remote sensing ecological index (*RSEI*) and ecological index (*EI*), and found that *RSEI* can reflect changes in ecological environment quality more effectively than *EI* (Shan *et al.* 2019). However, at present, most research based on *RSEI* can only evaluate the regional ecological quality by calculating the data of various indicators. There is no corresponding ecological evaluation platform that has been built, and this cannot be visualized in real time.

In the digital city stage, the main task is to build a digital spatial information-sharing platform. The platform design of this stage focuses on the sharing of basic spatial data. The platform design is based on a distributed spatial database, and the data provides online services through a Web service mode. With the development of smart city construction, the space–time data integration platform based on the data center, data warehouse or virtual view gradually appears, and finally develops into a space–time data cloud platform. However, due to the complex characteristics of space–time data, the space–time data cloud platform developed at present mainly focuses on technical implementation, while the research on the management, scheduling and visual analysis of massive space–time data is still in its initial stage. Cesium is a Javascript library that can display a 3D Earth and map across platforms and browsers. It supports 2D, 2.5D and 3D map expansion and dynamically displays data based on the timeline, providing good data visualization support (Birowosuto *et al.* 2007; De Paor *et al.* 2016). Therefore, the innovation of this study is to quantify the change system of the ecological environment in the research area from 2013 to 2018 for the first time and display its development process. So from the temporal characteristics of spatio-temporal data, this study adopts the open source web 3D map engine Cesium for development, fully considering remote sensing data, temperature, atmospheric data, soil data such as a data integration model, built with multi-source and heterogeneous data, the unity of the multi-dimensional space–time big data organization and management, and geographic space and time large-scale data calculation, a unified framework of time and space in visual display and the integrated service function of the ecological data integration processing platform, used to show changes in the ecological environment.

CONSTRUCTION OF ECOLOGICAL EVALUATION PLATFORM BASED ON CESIUM

Overall design

In order to realize the real-time extraction, query and analysis of ecological information, as well as the integration and decision analysis of massive real-time monitoring data, the holographic transformation of the ecosystem from static to dynamic is realized. In this study, a real-time rendering and interactive visualization network 3D platform based on the Cesium engine was constructed to realize the dynamic visualization of heterogeneous spatiotemporal data. Cesium supports dynamic geospatial visualization and has very high performance and precision. In view of this, this study chose Cesium as the basic engine of the network 3D data integration platform. The overall design of this module is shown in Figure 1. The platform has added a layer control module, geographic data management module, search query module, graph calculation module and coordinate positioning module. Based on the Cesium engine, user interaction functions such as *RSEI* index calculation, ecological decision evaluation and ecological environment evaluation can be realized. The whole platform adopts the Java language, Hadoop architecture for distributed integration computing, and Cesium engine for data integration visualization analysis in the front end.

The platform constructed in this study has improved the basic functions of Cesium, facilitating the rapid construction of the Cesium map project and its agile development, being reusable, supporting various configurations, and being suitable for ecological analysis. The platform integrates the leading open source to gallery, visualization, libraries, research and development of Web browser browsing using a three-dimensional Earth, without any plug-ins, provides a new big data visualization, real-time streaming data visualization function, through the platform can be realized fast on the browser and the mobile terminal is beautiful, smooth map plotting and spatial analysis. The core content of the platform construction is to build a Cesium-based network 3D data integration decision-making platform. Firstly, a unified organization and management module of space–time data is developed in the collaborative production of multi-source heterogeneous data to realize preliminary data integration. In the target matching module, the network structure and parameters of the SOM neural network are determined, and the mapping relationship between data and parameters is established to facilitate the continuous matching and fusion of data. Finally, in the data fusion module, a convenient interface for data editing is provided to realize efficient dynamic visualization and decision analysis.

Ecological evaluation decision function

The Fuxian Lake, ranging from 24°21′26″N to 24°38′06″N and 102°49′10″E to 102°57′26″E, is located in Yuxi city, the central area of the Yunnan Province. It is the largest deep-water freshwater lake in China and the largest lake at the source of the Pearl River, which is related to the Nanpanjiang river system (Li *et al.* 2017). The hills around Fuxian Lake are mainly distributed in the north and east, with shallow depressions in the middle and a river flowing from the eastern boundary. Fuxian Lake is a mid-narrow faulted lake formed under the influence of a tectonic basin. Its deepest point in the north reaches about 156.8 m and its maximum width reaches 11.5 km. The narrowest region is 3.2 km in the middle (Chen *et al.* 2019). More than 120 rivers flow into this lake, at least 56 of which are seasonal streams. The climatic conditions are the subtropical and semi-humid monsoon climate of Central Asia. Annual rainfall is about 943 mm, with 80%–90% concentrated in the monsoon season. The average temperature is about 15.5 °C (Cui *et al.* 2008; Zeng & Wu 2009). Seventy-five percent of the population live in the north of the Fuxian Lake area and the daily lives of the residents, cause local environment deterioration, therefore parts of northern basin range are included in the study area, in order to avoid the study area and nebula

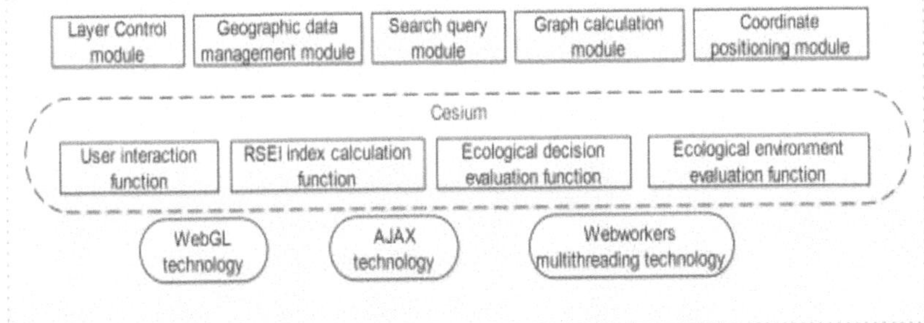

Figure 1 | Overall design of the Cesium module.

near Dianchi watershed river basin ecological environment affect the results of the study, according to the hydrological model based on DEM data extraction basin range. The Fuxian Lake Basin in the study area is shown in Figure 2.

To understand the evolution process of Fuxian Lake area's ecological environment, it is necessary to evaluate its ecological environment quality. Firstly, ecological security evaluation units should be determined, including vegetation coverage, water temperature, air humidity, soil dryness, heat and other indicators. To understand these environmental indicators, we need to get corresponding data and adopt appropriate methods to process them. In this study, Sentinel 2A and Landsat series remote sensing image data were used. The atmospheric environmental quality can be used, the humidity data calculated by the shore vegetation coverage and soil environmental quality and soil dryness heat index according to evaluation, and finally all the data entries are collected in the database.

Remote sensing technology is widely used in the land ecological security-related evaluation model, which has a specific way for realizing the quantitative description of ecological security. It can serve people's livelihood with the evaluation and analysis of regional ecological security, and put forward feasible Suggestions on ecological environment protection, governance and scientific development on the basis of evaluation. The remote sensing ecological index is a new ecological index used to comprehensively reflect the current situation of a regional ecological environment. The index is coupled with four indexes, namely, greenness, humidity, heat and dryness, which can be directly judged by human beings. It is completely based on remote sensing information technology, which reduces the subjectivity in practical application and the difficulty of index extraction. It can not only achieve quantitative evaluation of the ecological environment in the study area, but also visualize the evaluation results, and optimize the evaluation results to a certain extent. The technical flow chart of calculating the remote sensing ecological index is shown in Figure 3. In this study, Landsat8 data from 2013 to 2018 and Sentinel 2A remote sensing data from 2016 to 2018 were mainly used to control the cloud cover below 20%. Data preprocessing includes band synthesis, radiometric calibration and atmospheric correction, followed by calculation and extraction of the four indexes. The calculation results are normalized first, followed by principal component analysis, and finally the *RSEI* value is obtained.

Figure 2 | Schematic diagram of study area.

The normalized difference vegetation index (*NDVI*) is an important index used to detect plant growth status, vegetation coverage, leaf area index and other biochemical and physical properties of vegetation, and has been widely used in remote sensing monitoring of vegetation (Ren *et al.* 2017). The calculation of *NDVI* is shown in Equation (1), where *NIR* is near infrared band (band 8 in Sentinel data) and *R* is red band (band 4 in Sentinel data). The green index reflects the vegetation coverage of the study area. *Fv* is calculated by using the mixed pixel decomposition method, which roughly divides the landscape into water body, vegetation and buildings. The specific calculation formula is shown in Equation (2). In this study, $NDVIveg = 0.70$ and $NDVIsoil = 0.00$, and when *NDVI* of a pixel is greater than 0.70, the value of *Fv* is 1. When *NDVI* of a pixel is less than 0.00, the value of *Fv* is 0. Therefore, the vegetation coverage was used to replace the green index in this study.

$$NDVI = \frac{(NIR - R)}{(NIR + R)} \tag{1}$$

$$\begin{aligned} \mathrm{Fv} = {} & (\mathrm{NDVI\,GT}\ 0.7)*1 + (\mathrm{NDVI\,LT}\ 0.05)*0 \\ & + (\mathrm{NDVI\,GE}\ 0.05 \text{ and } \mathrm{NDVI\,LE}\ 0.7)* \\ & ((\mathrm{NDVI} - 0.05)/(0.7\text{–}0.05)) \end{aligned} \tag{2}$$

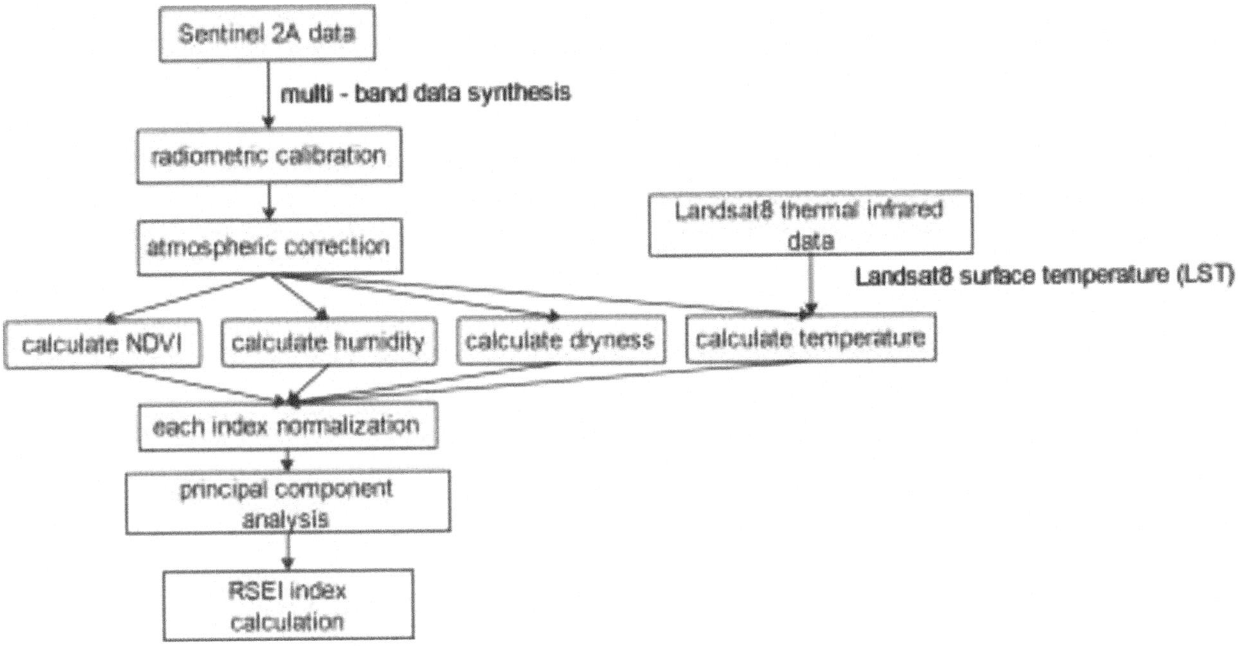

Figure 3 | Technical flow chart.

Taking remote sensing data in 2013 as an example, the green index diagram is shown in Figure 4(a). The components of brightness, greenness and humidity obtained by remote sensing hat transformation have been widely used in ecological environment monitoring. Among them, the component of humidity reflects the humidity of the water body, soil and vegetation and is closely related to the ecological environment. Therefore, the index of humidity is represented by the component of humidity. Due to the different types of sensors, the parameters vary accordingly. This research adopts the Tasselled Cap–Wetness formula published by the Index Database website, which

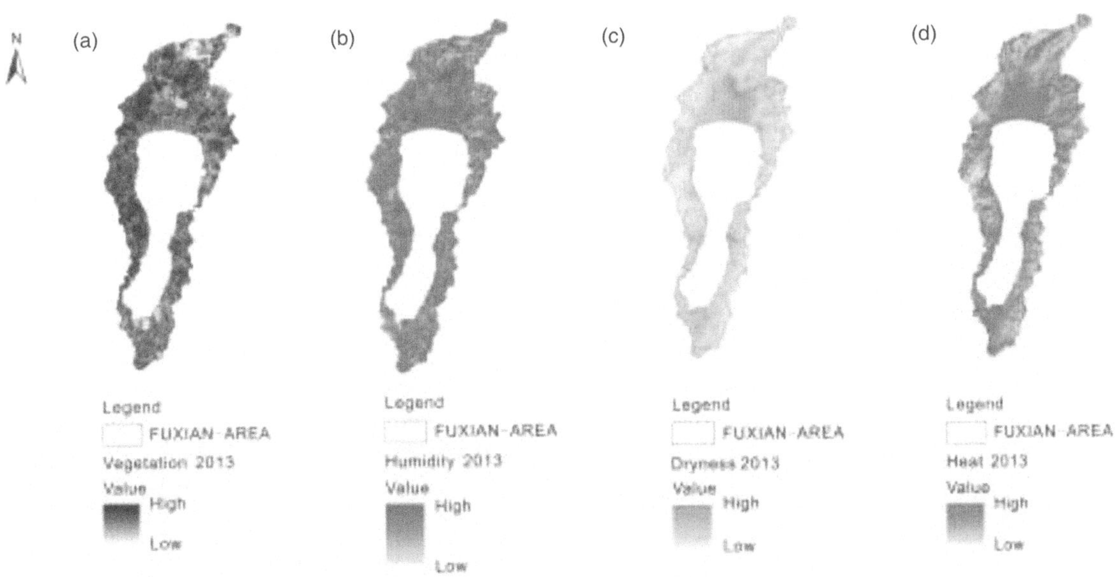

Figure 4 | (a) The green index diagram, (b) the humidity index diagram, (c) the dryness index diagram, (d) the heat index diagram).

calculates the moisture component of the Sentinel 2A data. As shown in Equation (3), each coefficient corresponds to the reflectance of each band of Sentinel 2A 2, 3, 4, 8, 11 and 12:

$$0.1509 * B2 + 0.1973 * B3 + 0.3279 * B4 + 0.3406 * B8 - 0.7112 * B11 - 0.4572 * B12 \quad (3)$$

Taking remote sensing data in 2013 as an example, the humidity index diagram is shown in Figure 4(b). The index of dryness is generally represented by the bare earth SI, but in the regional environment, there are quite a number of lands for construction, and they also cause the surface of the earth to be 'dry', so the dryness index can be made of a synthesis, namely the synthesis of bare soil index SI IBI Equation (4) and building index Equation (5); ρ_2, ρ_3, ρ_4, ρ_8, ρ_{11} are respectively the sentinel 2A data 2, 3, 4, 8, 11 band reflectance, and the dryness index (NDSI) is as shown in Equation (6), and the higher the value, the drier it is. Taking remote sensing data in 2013 as an example, the dryness index diagram is shown in Figure 4(c).

$$I = \frac{(\rho 11 + \rho 4) - (\rho 8 + \rho 2)}{(\rho 11 + \rho 4) + (\rho 8 + \rho 2)} \quad (4)$$

$$IBI = \frac{2\rho 11/(\rho 11 + \rho 8) - [\rho 8/(\rho 8 + \rho 4) + \rho 3/(\rho 3 + \rho 11)]}{2\rho 11/(\rho 11 + \rho 8) + [\rho 8/(\rho 8 + \rho 4) + \rho 3/(\rho 3 + \rho 11)]} \quad (5)$$

$$NDSI = \frac{(SI + IBI)}{2} \quad (6)$$

Finally, the Landsat8 surface inversion tool is used to represent the heat index with surface temperature (*LST*). Since Sentinel 2A data does not have thermal infrared bands, the thermal infrared bands of Landsat8 images acquired at the same time were used to calculate the temperature, and the temperature results obtained with a resolution of 30 m were resampled to 10 m, so as to match the Sentinel 2A 10 m spatial resolution and improve the discrimination degree of the surface thermal environment. Finally, Landsat8 TIRS data was used to calculate the surface temperature. In this study, the Landsat8 surface inversion tool based on an atmospheric correction method was directly used, and the surface temperature (*LST*) was used to represent the heat index (Zorer *et al.* 2011; Yao & Li 2019). By input using image information such as time, latitude and longitude, and model parameters, it and can calculate the band business, atmospheric transmission, effective bandpass upwelling radiance, effective bandpass downwelling radiance, and the resulting surface temperature traffic as the index of heat. Taking remote sensing data in 2013 as an example, the heat index diagram is shown in Figure 4(d).

APPLICATION EXAMPLE

Basic function

This study encapsulates the initialization of the map and initializes the map according to the configuration content by passing in JSON configuration information. In specific applications, some of the related parameters and layers are different in map initialization, and other functions are similar. According to need, different configuration information is modified or used. The configuration information is loaded as a JSON format file. The configuration file can be a static JSON file, and can also be loaded on the server dynamically according to user permissions to dynamically return data. Moreover, the platform supports the loading of almost all types of base maps in the current market, including ArcGIS services, Gaode, Google, world map and other online maps.

The basic functions of the platform constructed in this study include a query function, map configuration and layer control, measurement, coordinate positioning, marking, bookmark, map comparison, particle effect, flight tour, plotting, printing, etc. With specific functions such as the search function, by searching keywords it is possible to get place names with keywords and their location information. The function of measurement and plotting as shown in Figure 5 can calculate the spatial distance and ground distance (the distance measured according to the land's surface undulation, different from the mathematical linear distance in space) between the stored points or self-plotting points in the database, as well as the function of calculating area, angle, height, etc. The coordinate positioning function can locate the coordinate points on the map according to the user-input longitude, latitude and height information.

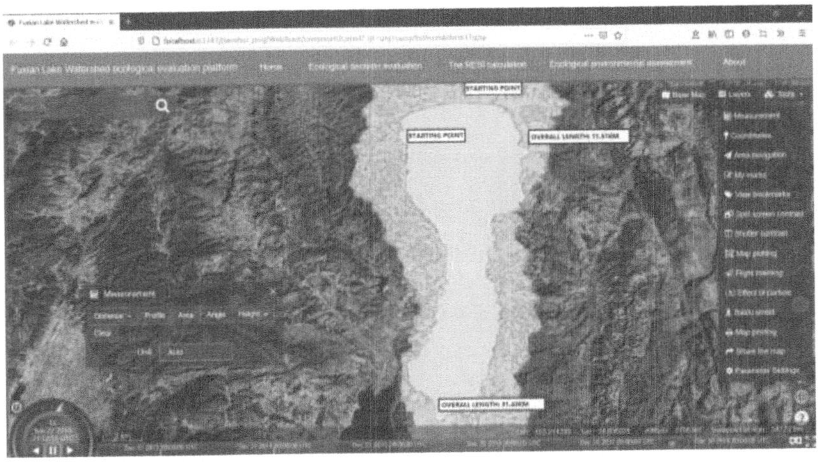

Figure 5 | Sample measurement function.

Map contrast features include layer split contrast and shutter contrast. The platform also builds a timeline, allowing users to view the ecological environment evaluation charts of different periods by dragging the timeline.

Ecological analysis function

Through the calculation in the section 'Ecological evaluation decision function' the four index values of green degree, humidity, heat and dryness can be obtained, and then the four indexes are normalized and analyzed by principal component analysis. Principal component analysis (PCA), based on the original data characteristics of the four indexes, automatically and objectively determines the weight value to realize the transformation of a variety of single indexes so as to be coupled into a comprehensive index, which avoids subjective arbitrariness, and the process is relatively simple. Because this study mainly analyzes the surrounding areas of Fuxian lake, the ecological environment change does not include the aquatic environment change. The load distribution on large bodies of water can affect PCA, and so in order to avoid water impact analysis results in the final analysis, the NDWI water index is used to mask information on the Fuxian Lake water body, and NDWI images are generated that highlight the waters (Puttinaovarat *et al.* 2015). Finally, the remote sensing ecological index was obtained. In this study, based on the calculated remote sensing ecological index, the ecological environment assessment analysis of the Fuxian Lake Basin in the long time series from 2013 to 2018 was carried out. The average calculation results of *RSEI* are shown in Table 1. The ecological environment of the study area continued to deteriorate from 2013 to 2015, improved slightly in 2016, and improved significantly in 2017–2018.

Table 1 | Calculation results of RSEI from 2013 to 2018

	2013	2014	2015	2016	2017	2018
RSEI	0.616	0.601	0.574	0.594	0.692	0.725

The *RSEI* result values of each grid were divided into five grades with equal intervals: poor (0.00–0.2), relatively poor (0.2–0.4), medium (0.4–0.6), good (0.6–0.8), and excellent (0.8–1.0), as the evaluation criteria for the ecological environment. As shown in Figure 6, the left side of the split screen shows the ecological evaluation results of the region in June 2013, while the right side of the split screen shows the ecological evaluation results of the region in June 2018. The green part shows good ecological environment, the yellow part shows medium ecological environment, and the red part shows poor ecological environment. It can be seen that in 2013, the overall ecological environment around the lake was medium, and the ecological environment around some residential areas was poor. In 2018, after a long period of treatment, the ecological environment of the vegetation-covered area around the lake was good, the ecological environment

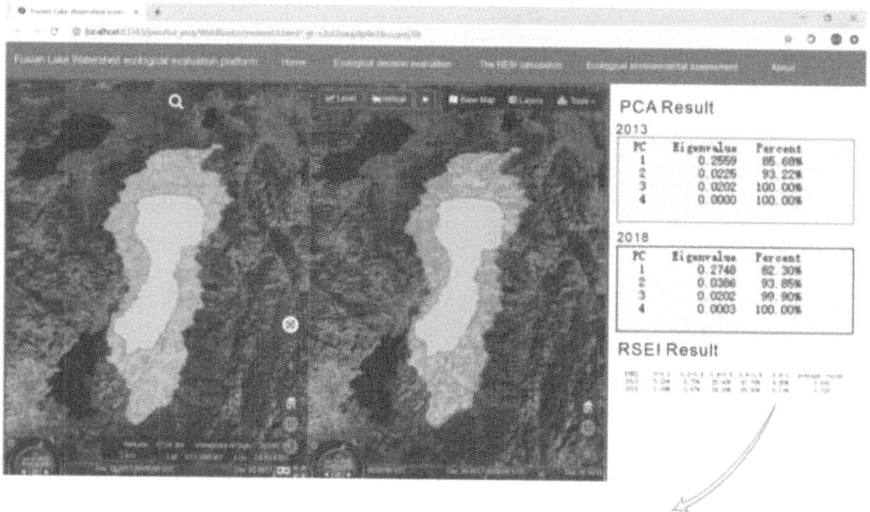

RSEI	0-0.2	0.2-0.4	0.4-0.6	0.6-0.8	0.8-1	average value
2013	5.21%	6.75%	35.43%	47.76%	4.85%	0.696
2018	1.04%	2.67%	24.36%	65.83%	6.10%	0.725

Figure 6 | Split screen comparison of ecological evaluation. Please refer to the online version of this paper to see this figure in colour: http://dx.doi.org/10.2166/ws.2020.253.

around the residential areas was medium, and the ecological environment of very few residential areas was poor. Since 2017, five major projects have been carried out in the Fuxian Lake Basin, including mountain forest expansion, land adjustment and water conservation, pollution control and river control, lake management and water conservation, and habitat restoration. Therefore, the results of ecological environment assessment are consistent with the actual situation, which also proves that the ecological protection and restoration project targeted at Fuxian Lake Basin is successful.

DISCUSSION AND CONCLUSION

In this study, a real-time rendering and interactive visualization network 3D platform based on the Cesium engine is constructed. The server can load a Google satellite map, DEM elevation map, and world map annotation. Based on Fuxian Lake Basin, this study can realize some real-time ecological analysis functions, which can display the change process of the ecological environment and natural resources, such as ecological environment quality evaluation, water color change, vegetation coverage area change and evaluation, etc. Search, navigation, distance measurement and other functions are realized. This design supports switching between different remote sensing data sources. Based on the existing geographic information data set, it contains an information window, which can conduct real-time ecological evaluation and display evaluation information, etc. According to the mean value of the RSEI index, this study quantitatively evaluates the ecological environment in the study area and presents the dynamic changes of the ecological environment over time. The research requires data to be standardized and weighted by principal component analysis. On the whole, the regional ecological environment quality with higher vegetation coverage rate is better, while the regional ecological environment quality that is greatly affected by human activities is worse. The results show that the ecological environment of the study area deteriorated from 2013 to 2015, improved slightly in 2016, and improved significantly in 2017–2018. The results are consistent with the actual situation and more accurate. For the first time, the change system of the ecological environment in the study area from 2013 to 2018 was quantified and its dynamics was displayed. The Cesium engine-based real-time rendering and interactive 3D visualization network platform was built, and also a multi-source, heterogeneous, multi-dimensional space–time unified organization with management of big data, geographical space and time large-scale data calculation, a unified framework of time and space of visual display and integrated service function of ecological data integration platform, which has realized the heterogeneous dynamic visualization and decision analysis of spatio-temporal data and is used to show the changes of ecological environment. However, this system can only analyze the existing data, and cannot realize the real-time monitoring function of any remote sensing data.

ACKNOWLEDGEMENTS

The authors would like to thank anonymous reviewers and the editors for their constructive comments and suggestions. This research was funded by the National Natural Science Foundation of China (41961064 and 41301520), the Scientific Research Foundation of Yunnan Province Education Department (2019Y0011) and the Yunnan Department of Science and Technology application of basic research project (Grant No. 202001BB050030).

DATA AVAILABILITY STATEMENT

All relevant data are included in the paper or its Supplementary Information.

REFERENCES

Birowosuto, M. D., Dorenbos, P., van Eijk, C. W. E., Krämer, K. W. & Güdel, H. U. 2007 Scintillation and luminescence properties of Ce^{3+} doped ternary cesium rare-earth halides. *Physica Status Solidi* **204** (3), 850–860.

Chen, J., Lyu, Y., Zhao, Z., Liu, H., Zhao, H. & Li, Z. 2019 Using the multidimensional synthesis methods with non-parameter test, multiple time scales analysis to assess water quality trend and its characteristics over the past 25 years in the Fuxian Lake, China. *Science of The Total Environment* **655**, 242–254.

Cui, Y. D., Liu, X. Q. & Wang, H. Z. 2008 Macrozoobenthic community of Fuxian Lake, the deepest lake of southwest China. *Limnologica* **38** (2), 116–125.

De Paor, D. G., Whitmeyer, S. J. & Bentley, C. 2016 Cesium – a virtual globe with strong potential applications in geoscience education. In: *51st Annual Northeastern GSA Section Meeting*.

Lintao, Z., Guofan, Y., Fuqiang, Z. & Juan, D. 2014 Ecological change assessment based on remote sensing index – a case study of Changning City. *Remote Sensing of Land Resources* **26** (4), 145–150.

Li, S., Zhou, J. & Wang, J. 2017 Spatial and temporal change and driving force analysis of land use/cover in Fuxian Lake basin from 1974 to 2014. *Remote Sensing of Land Resources* **29** (4), 132–139.

Mitrović, Đ. 2015 Broadband adoption, digital divide, and the global economic competitiveness of Western Balkan countries. *Ekonomski Anali* **60** (207), 95–115.

Puttinaovarat, S., Khaimook, K., Polnigongit, W. & Horkaew, P. 2015 Robust water body extraction from landsat imagery by using gradual assignment of water index and DSM. In: *IEEE International Conference on Signal & Image Processing Applications*, 19–21 October, Kuala Lumpur, Malaysia.

Ren, S., Chen, X. & An, S. 2017 Assessing plant senescence reflectance index-retrieved vegetation phenology and its spatiotemporal response to climate change in the inner Mongolian grassland. *International Journal of Biometeorology* **61** (4), 601–612.

Shan, W., Jin, X., Ren, J., Wang, Y., Xu, Z., Fan, Y., Gu, Z., Hong, C., Lin, J. & Zhou, Y. 2019 Ecological environment quality assessment based on remote sensing data for land consolidation. *Journal of Cleaner Production* **239**, 118126.

Song, H. M. & Xue, L. 2016 Dynamic monitoring and analysis of ecological environment quality in Weinan City based on remote sensing ecological index model. *Journal of Applied Ecology* **27** (12), 3913–3919.

Xu, H., Wang, M., Shi, T., Guan, H., Fang, C. & Lin, Z. J. 2018 Prediction of ecological effects of potential population and impervious surface increases using a remote sensing based ecological index (RSEI). *Ecological Indicators* **93**, 730–740.

Xu, H., Wang, Y., Guan, H., Shi, T. & Hu, X. 2019 Detecting ecological changes with a remote sensing based ecological index (RSEI) produced time series and change vector analysis. *Remote Sensing of Land Resources* **11**, 2345.

Yao, Y. F. & Li, L. I. 2019 Estimation of surface soil dryness based on landsat 8 TIRS data and temperature-vegetation dryness index. *Chinese Journal of Soil Science* **50** (2), 282–289.

Zeng, H. A. & Wu, J. L. 2009 Sedimentary records of heavy metal pollution in Fuxian Lake, Yunnan Province, China: intensity, history, and sources. *Pedosphere* **19**, 562–569.

Zorer, R., Rocchini, D., Delucchi, L., Zottele, F., Meggio, F. & Neteler, M. 2011 Use of multi-annual MODIS Land Surface Temperature data for the characterization of the heat requirements for grapevine varieties. In: *2011 6th International Workshop on the Analysis of Multi-Temporal Remote Sensing Images (Multi-Temp), Proceedings* (L. Bruzzone & F. Bovolo, eds), IEEE, Piscataway, NJ, USA, pp. 225–228.

First received 17 July 2020; accepted in revised form 26 September 2020. Available online 16 October 2020

Simulation of nitrogen and phosphorus pollution in typical agricultural and forested basins as well as relevant reduction effect based on SWAT model

Wu Li, Chen Zhe, Liu Hui-ying and Wang Dun-qiu

ABSTRACT

Non-point source nitrogen and phosphorus pollution is a critical threat to aquatic ecosystems and a potential risk to drinking water safety. To precisely control nitrogen and phosphorus pollution in the river basin, in this study, we identified key pollution areas of the river basin, analyzed the main characteristics of pollution sources as well as their contribution to the river basin pollution, and conducted simulation analysis on reduction measures based on the SWAT model. The results showed the following. (1) The simulation effect of the calibrated model was good, and sub-basins 3, 39 and 96 were the key source areas, the main sources of pollution were combined pollution from livestock and poultry breeding and planting industry. (2) Crops had the largest input and output for both nitrogen and phosphorus, the output of which was 6,137.8 t/a and 562.4 t/a respectively. The urban point sources had the highest output rates of nitrogen and phosphorus, 75.7% and 67.5% respectively. (3) With the optimal combination of reduction measures, nitrogen and phosphorus were reduced by 1,438.9 t and 85.3 t respectively, i.e., the reduction rates were the highest. The reduction effect for total nitrogen was better than that for total phosphorus.

Key words | agricultural and forested basins, nitrogen and phosphorus pollution, reduction, SWAT

Wu Li
Chen Zhe
Wang Dun-qiu (corresponding author)
Guangxi Key Laboratory of Environmental Pollution Control Theory and Technology, College of Environmental Science and Engineering, Guilin University of Technology, Guilin 541004, China
E-mail: *wangdunqiu@sohu.com*

Liu Hui-ying
Yongzhou Hydrology and Water Resources Investigation Bureau of Hunan Province, Yongzhou 425000, China

Wu Li
Environmental Protection Bureau of Yongzhou Economic and Technological Development Zone, Yongzhou 425000, China

HIGHLIGHTS

- Using local census data of pollution sources based on SWAT model.
- Identifying the key pollution sources through the characteristics of main pollution source and output.
- Using data of pollution sources examining contribution rate and output rate.
- The reduction effects of different reduction measures were compared.
- Using long-term hydrology, water quality data rate and validation, the result of model is reliable.

INTRODUCTION

The protection of water resources is essential for local social and economic development. To strengthen the protection of water resources in river basins, industrial point source pollution in major river basins of China has been effectively controlled. However, non-point source pollution is still a serious problem, which has significantly compromised water quality, and made the governance of river basins more difficult (Chen *et al.* 2016; Wang *et al.* 2019). The non-point source nitrogen and phosphorus pollutants are mainly from agricultural and urban pollution sources, among which the urban pollution sources include living pollution sources and industrial pollution sources in urban hubs and scattered rural areas. Agricultural pollution sources may be divided into livestock and poultry breeding, agricultural

doi: 10.2166/ws.2020.237

planting, and aquaculture, from which the time, route, quantity and category for pollution are uncertain. However, if their spatial distribution can be reflected accurately, the managers of river basins may accurately locate the key areas for governance. Therefore, some hydrological models, such as the HSPF model, AGNPS model and SWAT model, have been widely applied. The SWAT model takes into account soil type, land type, climate, and vegetation type, and its simulation results are characterized by strong temporal sequence and significant spatial distribution (Wang *et al.* 2020). Many scholars all over the world have studied regional non-point source pollution load and relevant distribution characteristics utilizing the hydrology–hydrodynamics–water quality linkage simulation function of the SWAT model, and accurately located the key source areas of relevant basins (Song *et al.* 2013; Chen *et al.* 2015). Analyzing the main characteristics of pollution sources and calculating the contribution rate of the pollution sources to the water quality pollution in relevant basins based on a comprehensive investigation of pollution sources may provide technical support for accurate formulation of policies concerning the basins. Previous studies have used output coefficients to calculate the input of nitrogen and phosphorus pollutants, so as to simulate the contribution and output rate of pollution sources, which often caused a great deviation from the actual situation (Mockler *et al.* 2017; He *et al.* 2019a, 2019b; Hua *et al.* 2019). Nitrogen or phosphorus pollution is closely related to rainfall runoff process, soil type, vegetation type, agricultural production mode (fertilizer application, tillage mode, etc.) and agricultural management measures. At present, many river basins have implemented different management measures, which have caused a great difference in nitrogen and phosphorus load reduction. Generally, the managers take the optimal integrated management measures dependent on local conditions, e.g. the optimal management measures for the Mary River Basin summarized by Richards *et al.* (2008), by which the phosphorus load was reduced by 50%. Wojciechowska *et al.* (2019) studied different management measures and policies in three small basins. In addition, for regions with serious soil and water loss, the prevention and control of soil and water loss and nutrient loss are often considered together in management measures for relevant basins, so as to make a comprehensive strategy (Niraula *et al.* 2012; Himanshu *et al.* 2019). In this study, a simulation of non-point source nitrogen and phosphorus pollution in the Yongzhou Basin of Xiangjiang River was carried out using local census data of pollution sources (from the second national census of pollution sources carried out recently) and farmland survey data (from the latest farmland quality survey report) based on the SWAT model, and three major goals for basin management and control were realized: (1) the key pollution areas of the river basin were identified; (2) the main characteristics of basin pollution sources and their contribution to the basin pollution were analyzed; (3) specific reduction measures were put forward, and the reduction effect was simulated. This study may be taken as a reference for effective control of nitrogen or phosphorus pollution in typical agricultural and forested basins, and especially the analysis of the presented key source areas, the apportionment of the sources in the basin and nitrogen and phosphorus pollution control countermeasures may support local initiatives better to protect the source of Xiangjiang River.

OVERVIEW OF THE STUDY REGION

Overview of the region

The study region is located on the upper reaches of the Xiangjiang River. Xiangjiang River is the largest river in Hunan Province, and a major tributary of the Yangtze River as well. It originates from Haiyang Mountain, Lingchuan County, Guangxi Zhuang Autonomous Region in the west, and Wild Dog Ridge, Lanshan County, Hunan Province in the east. Its total length is 948 km, and the area of the basin is 94,700 km^2. As a basin with the most abundant water resources and the highest development and utilization rates, it belongs to the Dongting Lake system and is dominated by a subtropical monsoon climate, with long-term average annual precipitation of 1,200–1,900 mm, and annual average temperature at 17.6–18.6 °C. Red soil and yellow soil are its main soil types, with perennial average sunshine between 101.5 and 113 $kcal/cm^2$.

Planting conditions

In this study region, double-season rice, maize, and oilseed rape are the main crops. To simulate soil nitrogen and phosphorus loss with farmland runoff, it is assumed that the double-season rice is planted in paddy fields and single-season maize in upland fields every year. The main fertilizers are nitrogen fertilizers and compound fertilizers. Urea with nitrogen content of 46% is the main nitrogen fertilizer while for the compound fertilizers, mainly a fertilizer containing nitrogen, phosphorus and potassium, 15% content for each, is adopted. For double-season rice and maize, there are base fertilizer and topdressing fertilizer for each season of rice. The application of fertilizers for the main crops is as shown in Table 1.

STUDY METHODS

Database construction

The database for the SWAT model was divided into two categories: spatial database and attribute database. All spatial data were converted into the projection coordinate system: WGS_1984_UTM_Zone_49N; after that the data of meteorological, land use type, soil type and the daily load of urban point sources and livestock and poultry breeding pollution sources were used for simulation, likewise the hydrology and water quality data were used for model calibration and verification respectively. See Table 2 for the names, parameters and sources of the main data.

Model construction and land use distribution

In this study, ARCSWAT was used to perform DEM image recognition, calculate the catchment area of the basin, extract the river network and delimit the sub-basin. Based on the thresholds of land use type, soil type and slope division set as 10%, 10% and 15%, the total basin area was 18,633.6 km^2, and the basin was divided into 115 sub-basins and 818 HURs. There were eight types of land in the study region. The land use types that covered the largest areas were forest land, paddy field and dry land, of which the areas were 11,393.5 km^2, 3,186.8 km^2 and 2,057.2 km^2 respectively, accounting for 89.3% of the total basin area of the study region together. The study region is a typical region dominated by agriculture and forestry, as shown in Figures 1 and 2.

Table 1 | Application of chemical fertilizers for the main crops

Category	Rice		Agricultural land
	Early rice	Late rice	Crop
Nitrogen fertilizer (kgN·ha^{-1})	10.8	9.5	10.6
Phosphate fertilizer (kgP·ha^{-1})	4.6	3.7	5.1

Soil type distribution

The soil type map reflects the spatial distribution of soil in the study region. This study utilized HWSD data downloaded from the FAO website to extract a 1:1,000,000 soil type map of Yongzhou, effective numerical range: 11,333–11,927. Some attribute data can be found through the HWSD_DATE database, and the rest of the data can be calculated by SPAW software. Thus, the soil database can be established (Figure 3).

MODEL CALIBRATION AND EVALUATION OF THE MODEL

Model calibration

In this study, the SUFI method of SWAT-CUP software was used for sensitivity analysis of model parameters, in which 71 parameters related to runoff and nitrogen or phosphorus cycle were selected for iterative modelling with the Global Sensitivity Analysis method. The parameters with the highest sensitivity are as shown in Table 3.

Evaluation of the model

In the study, the coefficient of determination (R^2) and Nash–Sutcliffe efficiency coefficient (Ens) were selected as the indexes for evaluation of the model. The coefficient of determination (R^2) reflects the correlation between simulated value and actual value of the model. The larger R^2, the better the consistency between the simulated value and the actual value. The Nash–Sutcliffe efficiency coefficient (Ens) represents the overall efficiency of model simulation.

Table 2 | Parameters required for SWAT model database establishment

Type		Format	Data type	Source	Remarks
Spatial data	Digital elevation model	GRID	Elevation data	Geospatial data sites	STRM with 12 m × 12 m resolution
	Land use type	GRID	Land use type	Resource and Environmental Science Center, Chinese Academy of Sciences	Land use map with resolution of 1 km × 1 km
	Soil type	GRID	Soil distribution type	Resource and Environmental Science Center, Chinese Academy of Sciences	Soil type map with resolution of 30 m × 30 m
Attribute data	Meteorological data	TXT table	Daily rainfall, the maximum and minimum mean temperature, daily radiation, wind speed, relative humidity, evaporation capacity, etc.	China Meteorological Data Network	Term of data for Dao County and Lengshuitan station: 2000–2019
	Hydrology and water quality data	EXCEL table	Monthly runoff, nitrogen and phosphorus concentrations	Yongzhou Hydrographic Bureau	Term of data: 2000–2019
	Pollution source data	EXCEL table	Data of pollution sources from urban point sources, livestock and poultry breeding, and planting industry	Yongzhou Ecological Environment Bureau	Data from the second national census of pollution sources
	Data on crops and fertilizers	Word file	Crop species, fertilization data, agricultural management measures	Yongzhou Agricultural and Rural Bureau	Farmland quality survey report
	Soil data		Hydrologic characteristics, density, electrical conductivity as well as physical and chemical properties of soil	Scientific Data Center for Cold and Arid Regions	HWSD soil texture data set
	Management measures		Planting pattern as well as fertilization dosage and time	Field survey and relevant statistical yearbook	

The closer the value to 1, the better the suitability of the model simulation. The formulas for calculation of R^2 and Ens are as follows:

$$R^2 = \left[\frac{\sum_{m=1}^{n}(O_m - \bar{O})(P_m - \bar{P})}{\sum_{m=1}^{n}(O_m - \bar{O})^2 \sum_{m=1}^{n}(P_m - \bar{P})^2}\right]^2$$

$$Ens = 1 - \frac{\sum_{m=1}^{n}(O_m - P_m)^2}{\sum_{m=1}^{n}(O_m - \bar{O})^2}$$

where O_m = the measured value for the mth time, $\bar{O}$ = the average measured value during the entire simulation period, P_m = the simulated value for the mth time, $\bar{P}$ = the average simulated value during the entire simulation period.

Usually, R^2 and Ens should be combined for evaluating the model parameter calibration. Generally speaking, as long as both $R^2 > 0.6$ and Ens > 0.5 are satisfied, the simulation results are acceptable (Moriasi *et al.* 2015). To avoid errors caused by the zero of many variables in the initial operation stage of the model, the first five years were selected as the warm-up period of the model, so as to improve the accuracy of the model. Spatially, the data from Dao County station and Lengshuitan station were selected for calibration in the upstream and downstream respectively, and the data were calibrated in sequence from the upstream to the downstream. The flow rate was calibrated first, and then the ammonia nitrogen and total phosphorus were calibrated (Zhao *et al.*2017).

Runoff, ammonia nitrogen and total phosphorus simulation: Dao County station took the years 2005–2017 as the period for calibration while Lengshuitan station took the years 2012–2017 as the period for calibration. At the same time, the years of 2018 and 2019 were taken as the

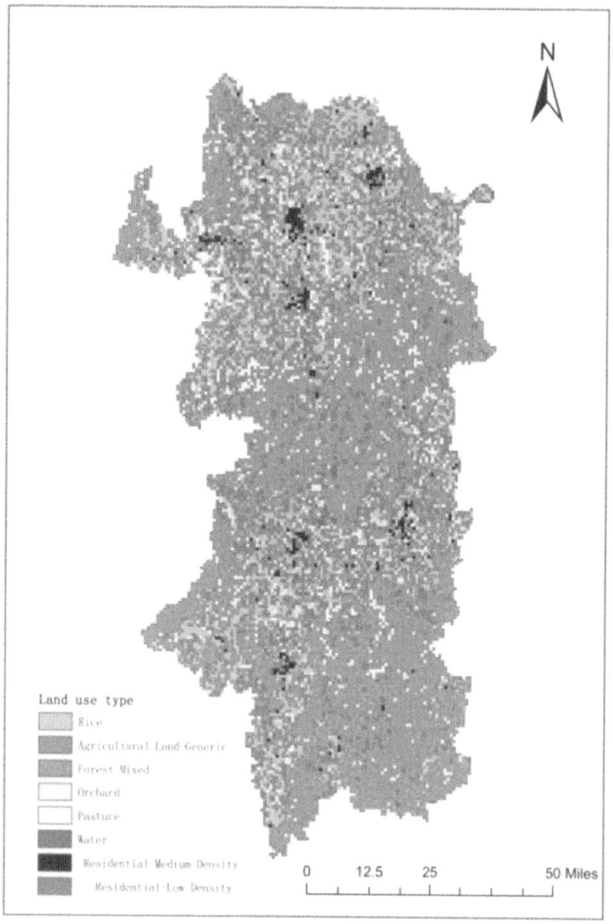

Figure 1 | Land use types in Yongzhou Basin of Xiangjiang River in 2018.

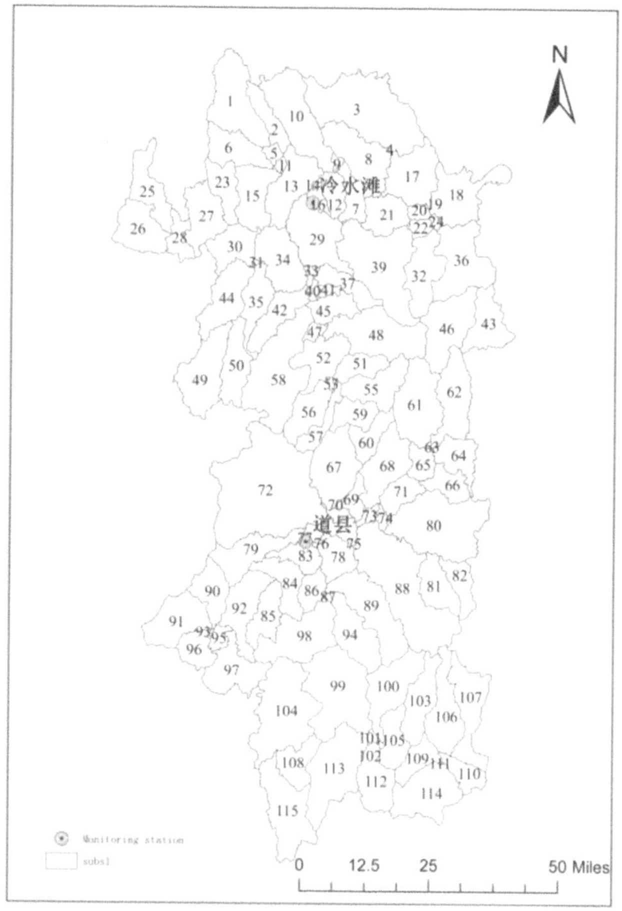

Figure 2 | Division and station distribution of Yongzhou Basin of Xiangjiang River in 2018.

validation period. For simulation accuracy values of runoff, ammonia nitrogen and total phosphorus of the two stations, see Figure 4. $R^2 > 0.68$ and Ens > 0.61 for runoff, ammonia nitrogen and total phosphorus of the two stations indicated that the SWAT model was suitable for Yongzhou Basin of Xiangjiang river, and Lengshuitan station is better than Dao County station in simulation accuracy.

RESULTS AND DISCUSSION

Analysis of characteristics of main pollution sources

There were a total of 363 urban point sources as the pollution sources in the study region, including 18 centralized sewage treatment plants in urban areas, 143 centralized sewage treatment facilities in rural areas (used for statistics of rural domestic sewage emissions), 202 industrial enterprises (with wastewater discharged directly into the river channel), and 2,758 livestock and poultry farms above the designated scale. In the region, the total nitrogen emission load was 12,927.12 t/a, the total phosphorus emission load was 1,501.82 t/a, the total nitrogen emission load and the total phosphorus emission load for the planting industry were 9,464.1 t/a and 864 t/a respectively, for livestock and poultry farms above the designated scale were 2,362.29 t/a and 575.66 t/a respectively, and for point sources in urban areas were 1100.73 t/a and 64.32 t/a respectively. The top six total nitrogen and phosphorus inputs from high to low in Yongzhou Basin of Xiangjiang River are as shown in Table 4 and Table 5. The maximum sub-basin inputs of total nitrogen and total phosphorus were 796.01 t and 88.49 t respectively, and the top three sub-basins for total

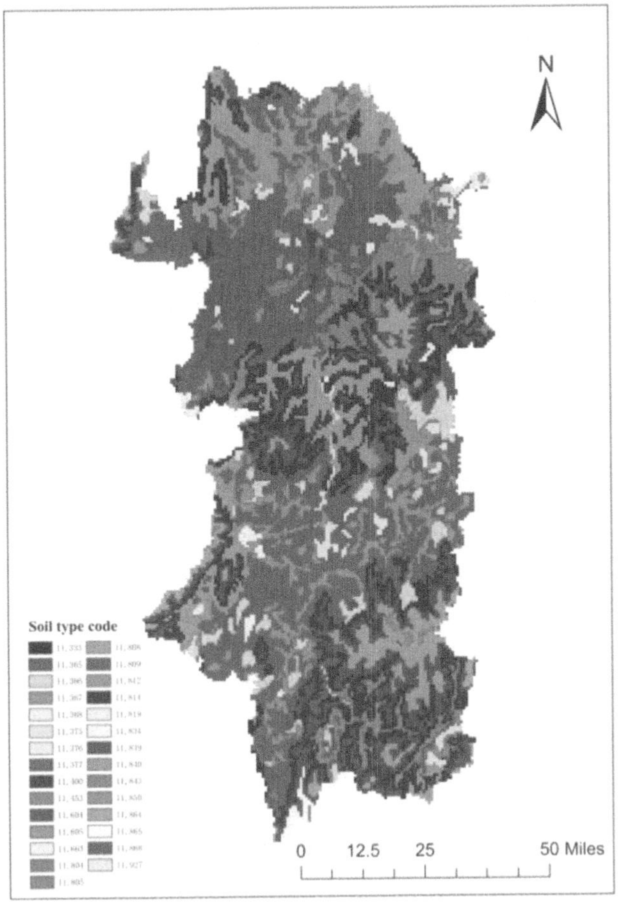

Figure 3 | Soil type map of Yongzhou Basin, Xiangjiang River in 2018.

nitrogen input were sub-basins 3, 72 and 39. Therefore, sub-basins 3 and 39 are the most centralized areas for output of total nitrogen and total phosphorus. The top three sub-basins for total phosphorus input were sub-basins 3, 39 and 104. Sub- basin 3 is located in the northwest of Qiyang County, including 13 urban point sources, two centralized sewage treatment plants and 143 livestock and poultry farms above the designated scale. For the planting industry, livestock and poultry breeding, and urban point sources, the total nitrogen input accounted for 68.49%, 19.89%, and 11.62% respectively, and the total phosphorus input accounted for 56.25%, 38.22%, and 5.54% respectively. The main sources of pollution are the combined pollution from the planting industry, livestock and poultry breeding, and urban point sources, mainly of the planting industry and livestock and poultry breeding. Sub-basin 39 is located in the northeast of Lingling District, including six urban point sources, and 80 livestock and poultry farms above the designated scale. For the planting industry, urban point sources, and livestock and poultry breeding, the total nitrogen emission load accounted for 74.52%, 0.11% and 25.37% respectively, and the total phosphorus emission load accounted for 49.28%, 0.27% and 50.45% respectively. The main sources of pollution were the combined pollution from the livestock and poultry breeding and the planting industry, mainly from livestock and poultry breeding.

Table 3 | Evaluation of model simulation results of the study region

Runoff parameter	Definition of parameter	N parameter	Definition of parameter	P parameter	Definition of parameter
ESCO	Soil evaporation compensation coefficient	GW_REVAPMN	Net flow coefficient of shallow groundwater	ERORCP	Phosphorus enrichment rate
CN2	SCS runoff curve number	ERORGN	Nitrogen permeability coefficient	BIOMIX	Biomixing efficiency coefficient
CN_K2	Effective hydraulic conductivity of main channel bed	CN2	SCS runoff curve number	PPERCO	Phosphorus infiltration coefficient
GW_REVAPMN	Net flow coefficient of shallow groundwater	SHALLST_N	Nitrate concentration in the basin discharged to the channel through groundwater	PHOSKD	Soil phosphorus distribution coefficient
SURLAG	Surface runoff lag coefficient	NPERCO	Nitrogen permeability coefficient	AI2	Proportion of P in the biomass of algae
ALPHA.BF	Base flow coefficient	AI6	Nitrogen portion in the biomass of algae	CH_OPCO	Organophosphate concentration in channel

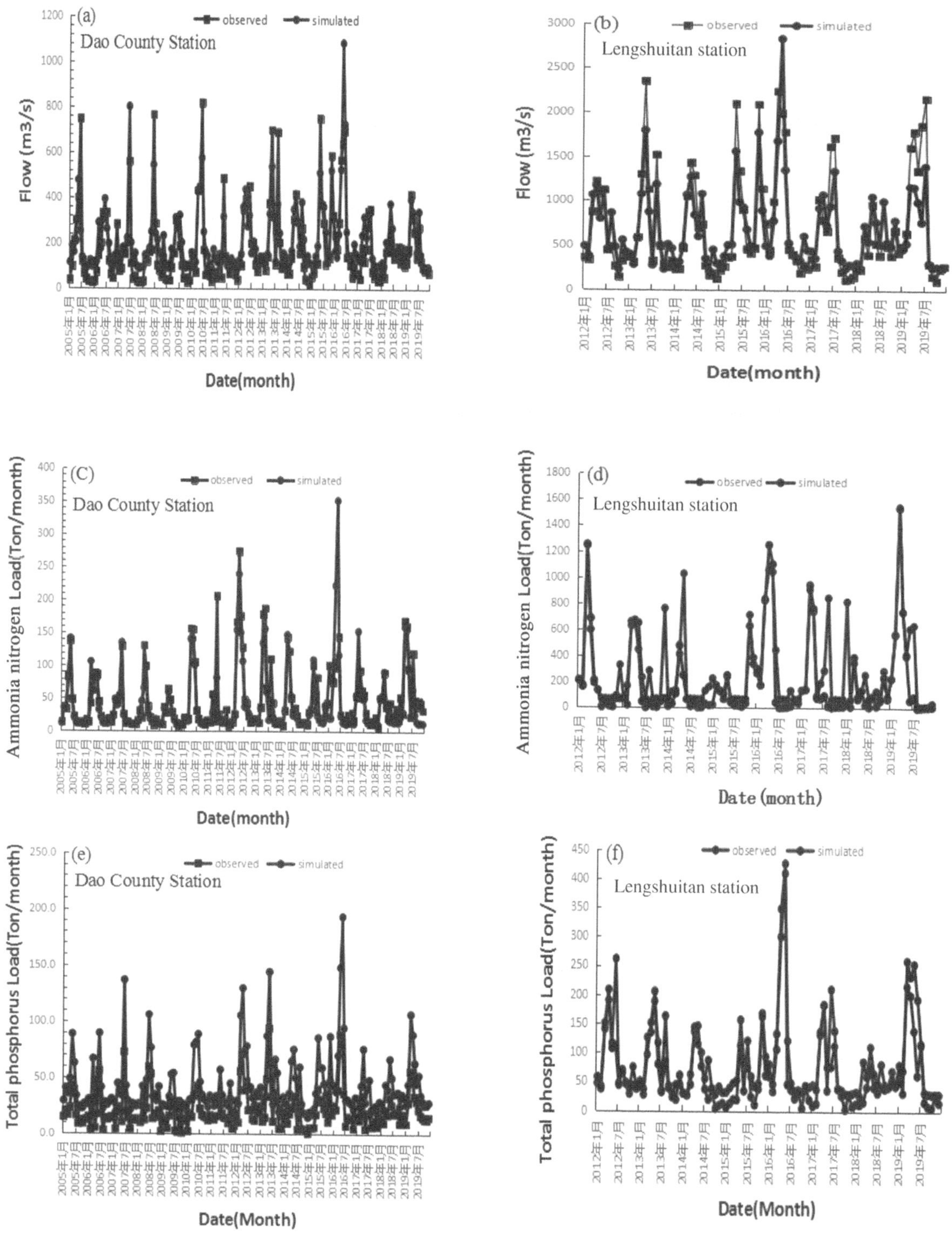

Figure 4 | Calibration and verification results of runoff, ammonia nitrogen and total phosphorus in Lengshuitan station and Dao County station.

Table 4 | Total nitrogen emission and its proportion in the Yongzhou Basin of Xiangjiang River

Sub-basin code	Crop cultivation/t	Point source in town/t	Livestock pollution/t	Total nitrogen emission/t	Proportion of crop cultivation(%)	Proportion of point source in town(%)	Proportion of livestock and poultry breeding (%)
3	545.20	92.46	158.35	796.01	68.49	11.62	19.89
29	293.60	0.53	98.26	392.39	74.82	0.14	25.04
39	360.26	0.53	122.63	483.41	74.52	0.11	25.37
72	454.30	1.32	67.96	523.57	86.77	0.25	12.98
80	298.97	125.27	56.69	480.93	62.16	26.05	11.79
104	259.33	4.74	140.46	404.53	64.11	1.17	34.72

Note: Crop-planting emissions in each sub-basin = emissions load for planting industry × farmland area/total area of sub-basin.

Table 5 | Total phosphorus emissions and proportion in the Yongzhou Basin of Xiangjiang River

Sub-basin code	Crop cultivation/t	Point source in town/t	Livestock pollution/t	Total phosphorus emission/t	Proportion of crop cultivation(%)	Proportion of point source in town (%)	Proportion of livestock and poultry breeding (%)
3	49.77	4.90	33.82	88.49	56.25	5.54	38.22
29	26.80	0.06	19.66	46.52	57.62	0.13	42.25
39	32.89	0.18	33.67	66.74	49.28	0.27	50.45
72	41.48	0.18	17.80	59.46	69.75	0.31	29.94
80	27.29	4.49	15.31	47.09	57.96	9.53	32.51
104	23.68	0.33	40.35	64.35	36.79	0.51	62.70

Note: Crop-planting emissions in each sub-basin = emissions load for planting industry × farmland area/total area of sub-basin.

Analysis of spatial distribution characteristics concerning output intensity of nitrogen and phosphorus pollution

Due to the fact that the census data of pollution sources were based on year 2017, the model output results select 2017 data for analysis. The spatial distribution concerning input and output intensity of total nitrogen and total phosphorus in the whole basin is as shown in Figure 5(a) and 5(b). The sub-basins with large pollution load input and output are mainly distributed in the main stream of Xiangjiang River and its north tributary. The output intensity of total nitrogen and total phosphorus reached 0.44–15.52 kg·ha^{-1} and 0.02–1.41 kg·ha^{-1} respectively. Sub-basins 98–103 in the upstream basin of Xiaoshui River were less polluted, the total nitrogen output intensity was less than 2.90 kg·ha^{-1}, and the total phosphorus output intensity was less than 0.28 kg·ha^{-1}. In the north basin, sub-basin 35 at the upper reaches of Xiangjiang River, sub-basins 21 and 39 at the lower reaches of Xiangjiang River, tributaries 2 and 3 of Luhong River, and sub-basin 96 in the southern basin were seriously polluted, with total nitrogen output intensity greater than 11.09 kg·ha^{-1}, and total phosphorus output intensity greater than 1.11 kg·ha^{-1}. Obviously, for sub-basins 3 and 39 in the north, both inputs and outputs of total nitrogen and total phosphorus were very high while for sub-basin 96 in the south, the key source area for nitrogen and phosphorus treatment in this basin, the outputs of total nitrogen and total phosphorus were very high. The reason why those sub-basins had very high outputs of total nitrogen and total phosphorus was that the livestock and planting industry developed rapidly, Moreover ineffective measures have been taken to reduce emissions by local government. For instance, Jianghua county (located around sub-basin 96) cooperated with Guangdong Wen's Food Group Co., Ltd, to introduce many large-scale pig breeding projects. According to the Statistical Yearbook, high-standard farmlands were completed in Yongzhou City every year.

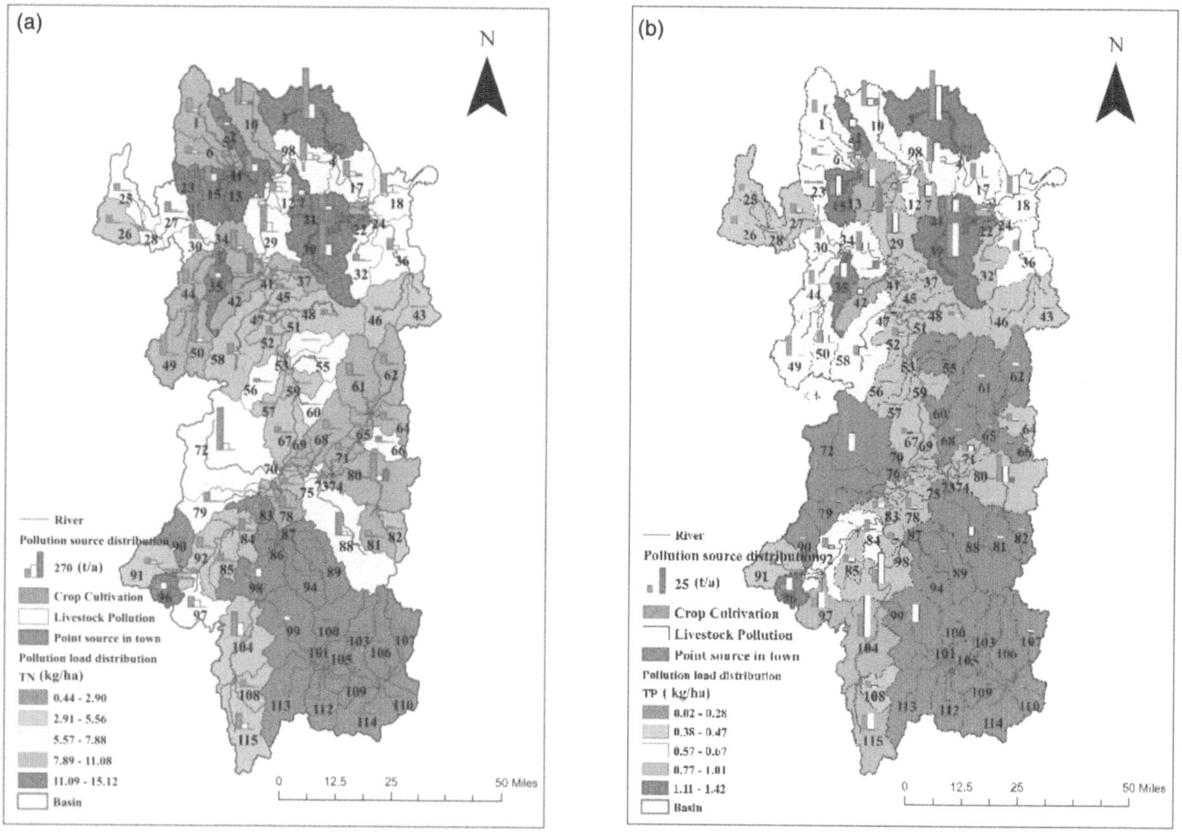

Figure 5 | (a) Spatial distribution of total nitrogen load in all basins in 2017. (b) Spatial distribution of total phosphorus load in all basins in 2017.

Analysis of output rate and contribution of nitrogen and phosphorus pollution

The pollution sources in the study region were divided into livestock and poultry breeding, crop cultivation and urban point sources. As shown in Table 6, for both total nitrogen input and total phosphorus input, crop cultivation ranked first (9,464.1 t/a and 864.0 t/a), livestock and poultry breeding ranked second (2,362.3 t/a and 575.7 t/a). For crop planting, livestock and poultry breeding and urban point sources, the pollution outputs (contribution rates) of the total nitrogen load model were 6,137.8 t/a (31.8%), 1,212.1 t/a (6.3%), and 833.5 t/a (4.3%) respectively, and of the total phosphorus load model were 562.4 t/a (25.5%), 242.4 t/a (11.0%), 43.4 t/a (2.0%) respectively, which were far lower than the background values and the values from other sources. The reason is that river background values and the contribution values from pollution sources of soil erosion and uncounted pollution sources were far greater than the obtained values. The total nitrogen output rates for pollution from crop planting, livestock and poultry breeding and urban point sources were 64.9%, 51.3% and 75.7% respectively, and the total phosphorus output rates for pollution from crop planting, livestock and poultry breeding and urban point sources were 65.1%, 42.1% and 67.5% respectively. Although the input of urban point sources was not high, the output rate was still high because of its direct discharge into the water body without soil and vegetation reduction, which was consistent with the results of Chen *et al.* (2015). The output rate of livestock and poultry breeding was the lowest. The reason may be that comprehensive utilization of agriculture reduced the output of relevant pollution sources (Zhang *et al.* 2014). The contribution rate of crop planting was the highest and the output rate of crop planting was also relatively high in the study region. The main reason for this was that the study region

Table 6 | Comparison table of total nitrogen and total phosphorus input and output of different pollution sources in 2017

	Input/t		Output/t		Output rate/%		Contribution rate/%	
	TN	TP	TN	TP	TN	TP	TN	TP
Background value and other sources	–	–	11,095.2	1,361.4	–	–	57.6	61.6
Crop source	9,464.1	864.0	6,137.8	562.4	64.9	65.1	31.8	25.5
Livestock and poultry breeding source	2,362.3	575.7	1,212.1	242.4	51.3	42.1	6.3	11.0
Town point source	1,100.7	64.3	833.5	43.4	75.7	67.5	4.3	2.0
Total	12,927.1	1,504.0	19,278.6	2,209.6	–	–	100	100

Note: Output rate = input/output; contribution rate = output of each pollution source/total output of pollution source.

was dominated by hills and mountains, where slope cultivation was prevalent and nutrient loss was caused by serious soil and water loss during rainfall (Zhao *et al.* 2017). Therefore, for this basin, the key for prevention and control is still fertilizer control in combination with soil erosion prevention and control, but joint control of livestock and poultry breeding and urban point sources is still necessary for reduction of nitrogen and phosphorus pollutants into the river.

Simulation of nitrogen and phosphorus pollution reduction

From 2020, global usage of chemical fertilizers will increase by 1.5% annually. However, the utilization rate of nitrogen and phosphorus fertilizers absorbed by crops is only 30%–50%. A large amount of leached nitrogen and phosphorus makes chemical fertilizers become the most important source of non-point source nitrogen and phosphorus pollution. Reducing chemical fertilizers is the fundamental measure for reduction of non-point source pollution (Tilman *et al.* 2002; FAO 2017). In terms of source control, the main reduction measure for livestock and poultry breeding is to use livestock and poultry manure as an organic fertilizer in the planting industry after being made hazard-free by treatment, which can not only reduce nitrogen and phosphorus pollutants directly discharged into the river, but also reduce the use of chemical fertilizers. This measure is also the key for nitrogen and phosphorus pollution control (Himanshu *et al.* 2019). According to a survey report concerning cultivated land quality in Yongzhou City of 2019, under the condition that the production increase of crops such as double-season rice and maize was guaranteed, the application amount of chemical fertilizers per mu was reduced by 2.47 kg/mu in 2018 as compared with that in 2019, and the application amount of chemical fertilizer will be reduced by 5% by 2020 through agricultural management measures, such as burying green manure, soil testing and formulated fertilization, returning straw to field, and replacing chemical fertilizers with organic fertilizers. At the same time, some studies showed that the application of resin-coated slow-release fertilizer and crop-specific fertilizer could reduce the usage of chemical fertilizers by 10%–25% without compromise of yield (Wu *et al.* 2002; Xu *et al.* 2002; Zhang *et al.* 2010). Therefore, this study set the fertilizer reduction rates at 5%, 10% and 20%. For comparison, the reduction rates for urban point sources and livestock and poultry breeding were set at 10% and 20% respectively. Because of the second three-year action plan for Xiangjiang River protection issued by the People's Government of Yongzhou City, the goal of the reduction rates for urban point sources and livestock and poultry breeding at 10% and 20% was feasible and practical. At present, emission reduction projects can be implemented such as by the retiring of livestock and poultry breeding in prohibited areas, building comprehensive utilization of manure in large-scale livestock and poultry farms and upgrading of sewage treatment facilities. As shown in Table 7, when the fertilizer reduction rate was set at 5%, the total nitrogen and total phosphorus of the basin were reduced by 288.4 t and 14.6 t respectively. Both reduction rates were greater than 20% for urban point sources and 10% for livestock and poultry breeding. When the fertilizer reduction rate was set at 20%, the total nitrogen and total phosphorus in the basin

Table 7 | Simulation of the reduction of total nitrogen and total phosphorus output under different reduction measures

	Urban point source emissions		Emissions from livestock and poultry breeding sources		Fertilizer application			The best combination of measures		
	Reduction 10% (A1)	Reduction20% (A2)	Reduction10% (B1)	Reduction 20% (B2)	Reduction 5% (C1)	Reduction10% (C2)	Reduction15% (C3)	Reduction20% (A1)	Reduction 20% (B2)	Reduction 15% (C3)
Total nitrogen reduction/t	16.2	82.1	103.1	210.6	288.4	596.5	1,146.2		1,438.9	
Total nitrogen reduction rate/%	0.1	0.4	0.5	1.1	1.5	3.1	6.0		7.5	
Total phosphorus reduction/t	1.8	5.4	12.8	26.3	14.6	28.9	53.6		85.3	
Total phosphorus reduction rate/%	0.1	0.2	0.6	1.2	0.7	1.3	2.4		3.9	

were reduced by 1,146.2 t and 53.6 t respectively, which were far greater than in urban point sources and livestock and poultry breeding of 20% reduction rate. The effect for total nitrogen reduction was better than for total phosphorus reduction, which was consistent with the results of Ma *et al.* (2016). The main reason why the reduction rate for livestock and poultry breeding was much lower than that for agricultural planting was that the pollution from livestock and poultry breeding below the designated scale was not included in the model in this study. Taking all the reduction measures into consideration, the optimal combination reduced total nitrogen and total phosphorus by 1,438.9 t and 85.3 t respectively. The larger the reduction rates for fertilizer application, livestock and poultry breeding and urban point sources, the lower the output of nitrogen and phosphorus pollution (Sun *et al.* 2013; Jang *et al.* 2016).

CONCLUSION

1. In the simulation period of Lengshuitan station and the calibration period and validation period of Dao County station, $R^2 > 0.68$ and Ens > 0.61 for runoff, ammonia nitrogen and total phosphorus. The simulation evaluation parameters met relevant requirements, which indicated that the SWAT model for Yongzhou Basin of Xiangjiang River had a good simulation effect.
2. The maximum inputs of total nitrogen and total phosphorus were 796.01 t and 88.49 t respectively; the output intensity of total nitrogen and total phosphorus reached 0.44–15.52 kg·ha^{-1} and 0.02–1.41 kg·ha^{-1} respectively; both sub-basins 39 at the lower reaches of Xiangjiang River and tributary 3 of Luhong River had total nitrogen output intensity greater than 11.09 kg· ha^{-1} and total phosphorus output intensity greater than 1.11 kg·ha^{-1}; as the key source areas for nitrogen and phosphorus treatment in this basin, both inputs and outputs of total nitrogen and total phosphorus were very high.
3. For both total nitrogen input and total phosphorus input, crop cultivation ranked first (9,464.1 t/a and 864.0 t/a), livestock and poultry breeding ranked second (2,362.3 t/a and 575.7 t/a). Crop planting had the highest pollution output and contribution rate of the total nitrogen load model, 6,137.8 t/a and 31.8% respectively; the nitrogen

and phosphorus output rates for pollution from urban point sources were the highest, 64.9% and 65.1% respectively, which indicated that chemical fertilizers are the most important source of non-point source pollution, and the application control of chemical fertilizers should also be combined with the control of livestock and poultry breeding and urban point source control.

4. In terms of source control, the reduction effect of chemical fertilizer application control was the best, and total nitrogen and total phosphorus could be reduced by 1,438.9 t and 85.3 t respectively, reduction rates: 7.5% and 3.9%. The effect for total nitrogen reduction was better than for total phosphorus reduction.

Nitrogen or phosphorus pollution is closely related to rainfall runoff process, soil type, vegetation type, agricultural production mode (fertilizer application, tillage mode, etc.) and agricultural management measures. Considering that terrace farming in typical agricultural and forestry basins is a common phenomenon, and soil erosion caused by terrace farming is closely related to water-body nitrogen and phosphorus pollution, it is therefore worthwhile to study the reduction effect of management measures such as medium-high planting, returning farmland to forest, and wetland buffer zone in the future. In addition, climate and land use patterns are also important factors affecting nitrogen and phosphorus pollution in the watershed.

FUND PROJECTS

National Natural Science Foundation of China (51638006); Guangxi Natural Science Foundation (2018GXNSFBA138039); Guilin University of Technology Research Startup Fund (GUTQDJJ2017015); Guangxi Key Laboratory of Environmental Pollution Control Theory and Technology Open Fund (Guikeneng 1701K006).

DATA AVAILABILITY STATEMENT

All relevant data are included in the paper or its Supplementary Information.

REFERENCES

Chen, D., Zhang, B., Zen, Y. F. & Zhang, H. P. 2015 Analysis of temporal and spatial characteristics of nitrogen pollution in Qingshan Lake basin based on SWAT model. *China Environmental Science* **35** (4), 1216–1222.

Chen, L., Wang, G. B., Zhong, Y. C. & Shen, Z. Y. 2016 Evaluating the impacts of soil data on hydrological and nonpoint source pollution prediction. *Science of The Total Environment* **563–564**, 19–28.

FAO 2017 *World Fertilizer Trends and Outlook to 2020: Summary Report*. Food and Agriculture Organization of the United Nations, Rome, Italy.

He, Q. W., Wendland, F. & Molkenthin, F. 2019a The analysis of nitrogen load and simulation uncertainty using SWAT in a catchment with paddy field in China. *Water Science & Technology* **80**, 806–816.

He, R. M., Yang, X. Y., Gassman, P. W., Wang, G. Q. & Yu, C. Y. Z. 2019b Spatiotemporal characterization of nutrient pollution source compositions in the Xiaohong River Basin, China. *Ecological Indicators* **107**, 105676.

Himanshu, S. K., Pandey, A., Yadav, B. & Gupta, A. 2019 Evaluation of best management practices for sediment and nutrient loss control using SWAT model. *Soil & Tillage Research* **192**, 42–58.

Hua, L. L., Li, W. C., Zhai, L. M., Yen, H., Lei, Q. L., Liu, H. B., Ren, T. Z., Xia, Y., Zhang, F. L. & Fan, X. P. 2019 An innovative approach to identifying agricultural pollution sources and loads by using nutrient export coefficients in watershed modeling. *Journal of Hydrology* **571**, 322–331.

Jang, S. S., Ahn, S. R. & Kim, S. J. 2016 Evaluation of executable best management practices in Haean highland agricultural catchment of South Korea using SWAT. *Agricultural Water Management* **180**, 224–234.

Ma, F., Jiang, X. F., Wang, L., Li, G. M. & Li, Z. 2016 Non-point source pollution control of Ashihe Basin based on SWAT model. *China Environmental Science* **36** (2), 610–618.

Mockler, E. M., Deakin, J., Archbold, M., Gill, L., Daly, D. & Bruen, M. 2017 Sources of nitrogen and phosphorus emissions to Irish rivers and coastal waters: estimates from a nutrient load apportionment framework. *Science of The Total Environment* **601–602**, 326–339.

Moriasi, D. N., Gitau, M. W., Pai, N. & Daggupati, P. 2015 Hydrologic and water quality models: performance measures and evaluation criteria. *Transactions of the ASABE* **58**, 1763–1785.

Niraula, R., Kalin, L., Wang, R. & Srivastava, P. 2012 Determining nutrient and sediment critical source areas with SWAT: effect of lumped calibration. *Transactions of the ASABE* **55** (1), 137–147.

Richards, C. E., Munster, C. L., Vietor, D. M., Arnold, J. G. & White, R. 2008 Assessment of a turfgrass sod best management practice on water quality in a suburban watershed. *Journal of Environmental Mannagement* **86** (1), 229–245.

Song, L. X., Liu, D. F., Xiao, S. B., Guo, H. C., Cui, Y. J. & Chen, L. 2013 Study on non-point nitrogen and phosphorus load from Xiangxi River in the Three Gorges Reservoir area based on SWAT. *Acta Scientiae Circurmstantiae* **33** (1), 267–275.

Sun, J. H., Zhu, Q. D., Lian, X. J., Qing, F. & Chen, C. 2013 Simulation analysis and best management practices for non-point source pollution in polder area of water net plan. *Resources and Protection of the Yangtze River Basin* **22** (Z1), 75–82.

Tilman, D., Cassman, K. G., Matson, P. A., Naylor, R. & Polasky, S. 2002 Agricultural sustainability and intensive production practices. *Nature* **418** (6898), 671–677.

Wang, G. Q., Li, J. W., Sun, W. C., Xue, B. L., Yinglan, A. & Liu, T. X. 2019 Non-point source pollution risks in a drinking water protection zone based on remote sensing data embedded within a nutrient budget model. *Water Research* **157**, 238–246.

Wang, K., Wang, P. X., Zhang, R. D. & Lin, Z. B. 2020 Determination of spatiotemporal characteristics of agricultural non-point source pollution of river basins using the dynamic time warping distance. *Journal of Hydrology* **583**, 124303.

Wojciechowska, E., Pietrzak, S., Matej-Łukowicz, K., Nawrot, N., Zima, P., Kalinowska, D., Wielgat, P., Obarska-Pempkowiak, H., Gajewska, M., Dembska, G., Jasiński, P., Pazikowska-Sapota, G., Galer-Tatarowicz, K. & Dzierzbicka-Głowacka, L. 2019 Nutrient loss from three small-size watersheds in the southern Baltic Sea in relation to agricultural practices and policy. *Journal of Environmental Management* **252**, 109637.

Wu, Z. J., Zou, Z. R., Cheng, R. F. & Chen, C. W. 2002 Effect of polythene-coated urea on greenhouse tomato. *Yangtze River Vegetables* **7**, 35–38.

Xu, P. Z., Zhang, F. B., Tang, S. H., Chen, J. S., Xie, C. S. & Chen, R. C. 2002 Preliminary report on application effect of special long-acting controlled-release BB fertilizer for rice. *Guangdong Agricultural Sciences* **1**, 18–20.

Zhang, S. J., Wang, L., Ma, F., Li, S. Y., Zhang, X. & Wu, J. T. 2010 Application of arbuscular mycorrhiza on promoting the growth of rice and reducing the usage of chemical fertilizer. *Journal of Harbin Institute of Technology* **42** (6), 958–962.

Zhang, H. L., Qi, S., Lu, Q. Q., Yang, B., Sun, J. & Liu, X. Y. 2014 Control zone of agricultural non-point source pollution based on water function zone in Hubei province. *Science of Soil and Water Conservation* **12** (2), 1–8.

Zhao, K., Su, B. L., Shen, M. M., Guan, Y. T. & Zhou, J. W. 2017 An improved method for parameter identification of SWAT model. *South-to-North Water Transfers and Water Science & Technology* **15** (4), 49–53.

First received 28 July 2020; accepted in revised form 16 September 2020. Available online 28 September 2020

Water quality assessment of Middle Route of South-North Water Diversion Project based on modified Nemerow index method

Yilin Yang, Dawei Zhang, Jin Quan, Pengwen Wang and Yi Xu

ABSTRACT

It is of great significance to find a scientific way to assess the water quality of the Middle Route of the South-North Water Diversion Project. In this paper, the spatio-temporal changes of the 26 water quality indicators in eight key monitoring stations along the Middle Route since the project was put into use were analyzed, and the modified Nemerow index method was employed to assess the water quality of the route in its early stage of operation. The results show that the water quality remained good in the period of study. Important water quality indicators were identified, including temperature, permanganate index, dissolved oxygen, total phosphorus, total nitrogen, mercury content, and density of planktonic algae. It is advised that potential ecological risks related to algae and shellfish reproduction in the open canal deserve special attention.

Key words | Nemerow index, South-North Water Diversion Project, water quality assessment, water quality index

Yilin Yang
Pengwen Wang
State Key Laboratory of Hydraulic Engineering Simulation and Safety,
Tianjin University,
Tianjin 300350,
China

Yilin Yang
Dawei Zhang
Jin Quan (corresponding author)
Yi Xu
China Institute of Water Resources and Hydropower Research,
Beijing 100038,
China
E-mail: *jeanquan@163.com*

Yi Xu
College of Civil Engineering and Architecture,
Zhejiang University,
Hangzhou 310058, Zhejiang province,
China

HIGHLIGHTS

- The water quality of the MR-SNWTP since the operation was analyzed.
- The modified Nemerow index method was applied to the South-North Water Diversion Project.
- The performance of the single factor method and the modified Nemerow index method was compared.
- The water quality indicators that deserve special attention were pointed out.
- The suggestion of strengthening monitoring of potential risks posed by algae was put forward.

INTRODUCTION

The South-North Water Diversion Project of China is a grand strategic infrastructure project designed to alleviate the water shortage in northern China, optimize allocation of water resources, guarantee sustainable socio-economic development, and build a well-off society in an all-round way (Wei *et al.* 2010; Tang *et al.* 2014). Since the beginning of the construction of the Middle Route of the South-North Water Diversion Project (MR-SNWDP), its water quality has been an issue of top concern around the nation, and the water quality standards are high and difficult to meet (Zhao *et al.* 2019). Meanwhile, as the society moves forward and urbanization advances, the water quality of the diversion project draws increasing attention. Especially in recent years, there are frequent reports about seasonal algae outbreaks and siltation in the canal (Lin *et al.* 2017; Zhu *et al.* 2019), which not only poses threat to the water quality and water supply safety of the project, but undermines sustainable development of the water-receiving areas.

doi: 10.2166/ws.2021.006

Water quality assessment is a process of selecting proper water quality indicators, water quality standards and calculation methods according to the assessment objective to assess the objective state of water quality by means of numerical values and images and thereby demonstrate the degree of water pollution (Veettil & Mishra 2016; Wu *et al.* 2017). In 1970, R. M. Brown *et al.* put forward the Water Quality Index (WQI) and N.L Nemerow put forward the Nemerow index method to assess the surface water pollution in the New York state (Angelakis *et al.* 1999; Liu *et al.* 2018; Long *et al.* 2019); recent years have seen the application of the fuzzy set theory, grey system theory and mathematical statistics theory to water quality assessment (Shi *et al.* 2017; Wang *et al.* 2017), but the widely discussed topics in water quality assessment research are mainly on the following three aspects: (1) studies on the concentration of selected water quality indicators are based on chemical or biological detection techniques; (2) water quality risk analysis is taken into account in water quality assessment; (3) studies on water quality assessment models. There is currently little research on open-canal water diversion projects under the artificial-natural complex conditions. To fill the research gap, this study intends to: (1) analyze and assess the water quality change characteristics of the main canal of the Phase I project of the MR-SNWDP (hereinafter referred to as the 'main canal') since the canal was put into use; (2) employ the modified Nemerow index method to perform a comprehensive assessment according to the water quality characteristics of the main canal and the realities of the project; (3) select and analyze key water quality indicators and offer scientific suggestions for project management in combination with the characteristics of the main canal of the Middle Route.

MATERIALS AND METHODS

Overview of the study area

The South-North Water Diversion Project is by far the largest and most important inter-basin water diversion project in China, consisting of the East Route, the Middle Route, and the West Route (Figure 1). The Middle Route project (32.67–39.98° N; 111.71–116.27° E) is a manually-dug open canal with a total length of 1,432 km that starts from the Danjiangkou Reservoir of Hubei, spans the Yangtze River, Huaihe River, Yellow River and Haihe River from south to north, and crosses Henan, Hebei, Tianjin and Beijing, diverting water to the four aforementioned provinces and municipalities directly under the central

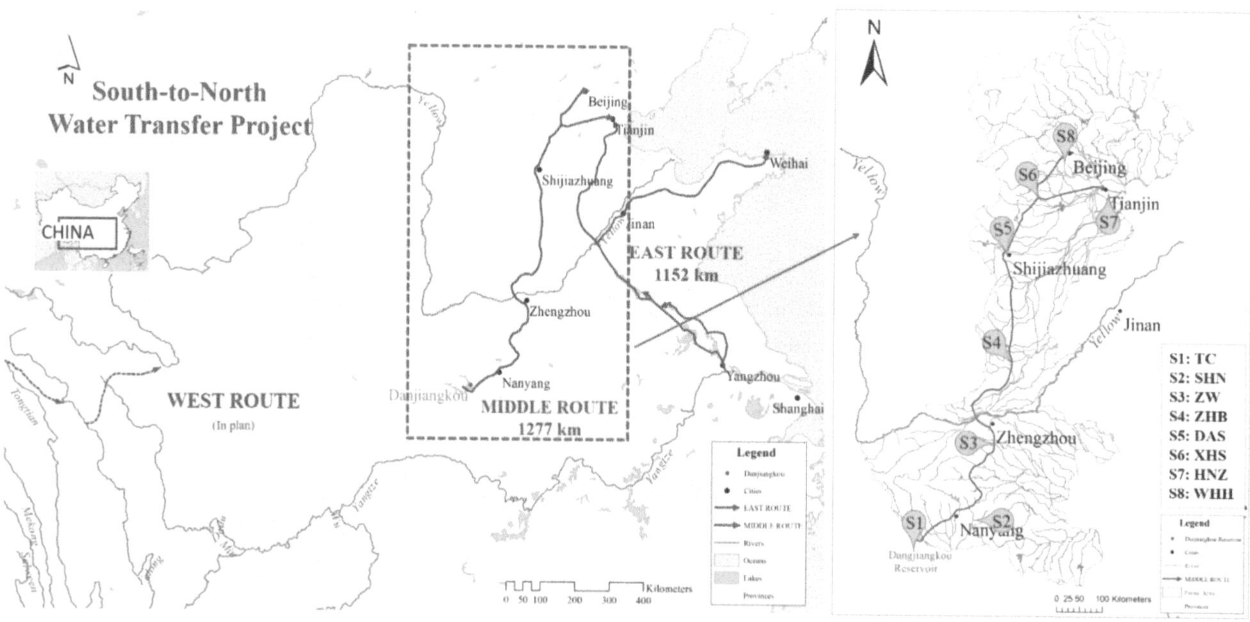

Figure 1 | Schematic diagram of the study area and the locations of the water quality monitoring stations.

government. Since the canal started to divert water on December 12th, 2014, the Middle Route has benefited about 60 million residents along the route.

Data source and sample test methods

The original water quality data used in this study are from the Construction and Administration Bureau of South-to-North Water Diversion Middle Route Project (the MR Administration Bureau) and official permission has been obtained for the analysis and processing of these data. Data from eight monitoring stations (Figure 1) were used in this study, and 26 water quality indicators were used, of which 24 are the routine test items stipulated in the Environmental Quality Standards for Surface Water of the People's Republic of China, and the remaining two are the monitoring items added by the MR Administration Bureau to its actual water quality management; the threshold standards were determined according to the regular items in the *Environmental Quality Standards for Surface Water* (GB3838-2002) and the standards specified in the *Water Quality Guide*. The sampling methods and the methods for chemical analysis of samples were based on the *Standards and Methods for Water and Wastewater Inspection* (Appendix A).

Establishment of assessment methods

Building the assessment indicator system

The water quality assessment indicator system of the main canal of MR-SNWDP was built based on data of normal operation of the long-distance water diversion project and statistics from the MR Administration Bureau. The characteristics and risks of the MR-SNWDP were comprehensively considered, and new indicators including bio-toxicity and ecological indicators were added to the system, as shown in Table 1.

Data processing

The indicators were not comparable due to the different dimensions and magnitudes of the original data, and normalization was thus required to exclude the influences of different dimensions and magnitudes (Long *et al.* 2019). In particular, the larger the index value of a positive indicator, the better; while the smaller the index value of a reverse indicator, the better. Here, the 'Range 0 ~ 1' method was used to process the data (Yang *et al.* 2020): suppose the indicators of the system to be $X_{ij} = (x_{ij})$ $(i = 1, 2, \ldots, n; j = 1, 2, \ldots, m)$. Normalization processing is performed on X_{ij}, and the normalization formula is as follows:

$$x_i' = \begin{cases} (x_i - x_{min})/(x_{max} - x_{min}), & x_i \text{ is a positive indicator; the greater } x_i \text{ is, the better} \\ (x_{max} - x_i)/(x_{max} - x_{min}), & x_i \text{ is a negative indicator; the smaller } x_i \text{ is, the better} \end{cases}$$

where x_{min} and x_{max} are the miminum and the maximum values of the indicator x_i.

RESULTS AND DISCUSSION

Water quality characteristic analysis

General indicators

The indicators of the assessment system were analyzed first, and then a further characteristic analysis was performed on indicators with obvious changes, including the pH value, water temperature (T), permanganate index (PI), five-day biochemical oxygen demand (BOD_5) and dissolved oxygen (DO). Among nutritional salts, total phosphorus (TP), total nitrogen (TN) and ammonia nitrogen (NH_3-N) were selected as the key indicators for in-depth analysis.

The annual average pH scale was between 8.1 and 8.4, indicating the water quality in the main canal of the MR-SNWDP remained alkaline throughout the monitoring and assessment period (Figure 2(a)). Water temperature (*T*) showed a trend of overall decline from south to north, in conformity with the temperature change pattern of China,

Table 1 | Water quality assessment indicator system for the main canal of the MR-SNWDP

Target layer	Primary criterion layer	Secondary criterion layer	Factor layer	Indicator attribute
Water quality assessment of the main canal of the MR-SNWDP	Conventional indicators	General indicators	Water temperature	Variable-amplitude indicator
			pH value	Range indicator
			Dissolved oxygen	Positive indicator
			Permanganate index	Reverse indicator
			Chemical oxygen demand (COD)	Reverse indicator
			Five-day biochemical oxygen demand (BOD_5)	Reverse indicator
			Ammonia nitrogen (NH_3-N)	Reverse indicator
			TP (measured in P)	Reverse indicator
			TN (measured in N in lakes and reservoirs)	Reverse indicator
		Toxicity indicators	Copper	Reverse indicator
			Zinc	Reverse indicator
			Fluoride (measured in F^-)	Reverse indicator
			Selenium	Reverse indicator
			Arsenic	Reverse indicator
			Mercury	Reverse indicator
			Cadmium	Reverse indicator
			Chromium (hexavalent)	Reverse indicator
			Lead	Reverse indicator
			Cyanide	Reverse indicator
			Volatile phenol	Reverse indicator
			Petroleum	Reverse indicator
			Anionic surfactant	Reverse indicator
			Sulfide	Reverse indicator
			Fecal coliform (ind./L)	Reverse indicator
	Unconventional indicators	Biotoxicity	Comprehensive biotoxicity	Reverse indicator
		Ecological indicators	Density of planktonic algae (DPa × 10^4 ind./L)	Reverse indicator

as shown in Figure 2(b). The maximum and minimum values were recorded at 32 °C and 0.5 °C, respectively, and the average temperature was 16–18 °C, without significant spatial differences. The PI value displayed a gradual increase from south to north along the canal, as shown in Figure 2(c). According to the limit of 2.0 mg/L for national standard (GB3838-2002) Level I water, only 50% of the sections met the standard, and it thus can be seen that the significant change trend of PI requires close attention. The BOD_5 of all the sections met the standard of GB Level I water and was less than 3.0 mg/L, as shown in Figure 2(d). COD and BOD_5 are indicators for the degree of pollution of organic matter to water. This showed the concentration of organic matter in the water along the canal was in a significant downward trend, and the possible cause is that the organic matter was utilized and assimilated by the animals and plants in the water (Cheng *et al.* 2019) or that the organic matter settled on a small scale with the utilization of sediment microorganisms and the like in the water (Shao *et al.* 2014). To sum up, the changes in the two indicators, PI and BOD_5, deserve special attention.

Dissolved oxygen (DO) is a positive indicator, and the average value at each station were all higher than 7.5 mg/L and met the requirements of GB Level I water, as shown in Figure 3(a). The changes in TP are presented in Figure 3(b), and the concentration was around 0.01 mg/L during the monitoring and assessment period, meeting the standard of Level I water. TN exhibited no obvious spatial differences and temporal changes from south to north, as shown in Figure 3(c). The annual average concentration

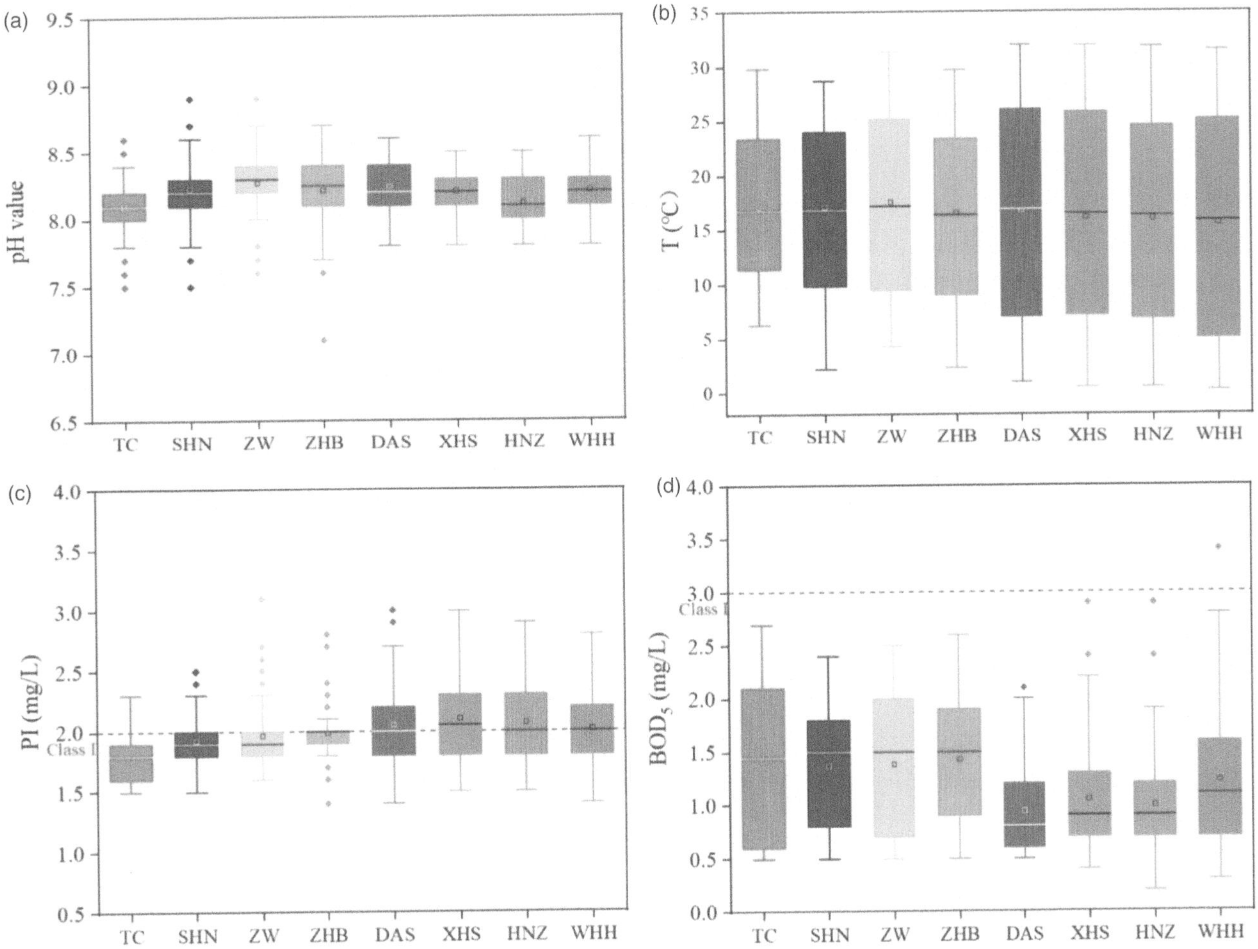

Figure 2 | Average concentration and standard deviation of (a) pH, (b) T, (c) PI and (d) BOD_5 in eight water quality monitoring sections of the MR-SNWDP from 2015 to 2018.

was around 1.0 mg/L, and the concentration at some sections was 50% higher than this concentration standard, which might be related to the high background value of total nitrogen in water source (Tan *et al.* 2015). Ammonia nitrogen did not show obvious changes from south to north, as shown in Figure 3(d). However, the ammonia nitrogen value exceeded the Level I water standard of 0.15 mg/L in a few stations in the latter stage, so it should be monitored as an important water quality indicator.

Toxicity indicators

Heavy metal ions and other toxic substances are important indicators affecting water quality. After comprehensive analysis of all toxicity indicators, mercury and arsenic were selected for in-depth analysis. The limit value of mercury in Level I water was lower than 0.00005 mg/L, and the values from all the sections met the requirement of Level I water, as shown in Figure 4(a); the values at several time points were lower than the detection limit of 0.0001 mg/L, but it requires long-term attention due to its high toxicity. The concentration of arsenic was all lower than the standard limit of Level I water, as shown in Figure 4(b). Arsenic showed a random trend and this might be related to the weathering and settlement of the rock mass or the soil composition in the vicinity (Liu *et al.* 2016). It is advisable to focus on the changes of these two indicators in future work and perform cumulative monitoring and assessment.

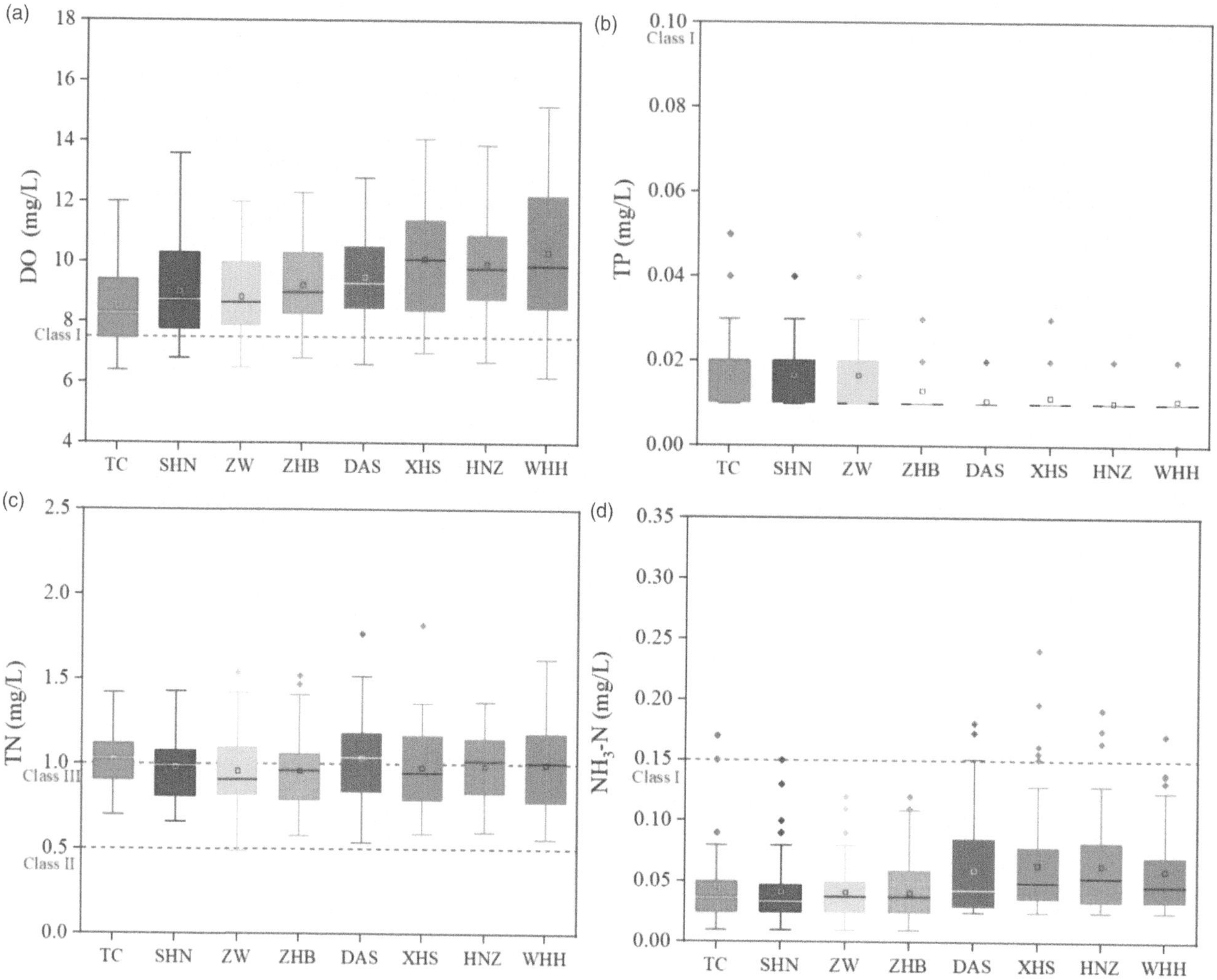

Figure 3 | Average concentration and standard deviation of (a) DO, (b) TP, (c) TN and (d) NH_3-N in eight water quality monitoring sections of the MR-SNWDP from 2015 to 2018.

Ecological factor indicators

The ecological factors for assessment include comprehensive biotoxicity, fecal *Escherichia coli* and density of planktonic algae. The percentage concentration of luminescent bacteria was used to monitor the comprehensive biotoxicity, and there was no case where the concentration of luminescent bacteria was higher than 30%, but the comprehensive biotoxicity remains a very important indicator. The overall average concentration of fecal *E. coli* in all monitoring stations was lower than 200 ind./L, meeting the requirements for Level I in GB, as shown in Figure 4(c). The concentration tended to increase gradually from south to north along the canal, and this increasing trend is particularly prominent in the northern part of the water-receiving area. Algae is another important ecological indicator. The annual average data in Figure 4(d) revealed that the density of planktonic algae (DPa) gradually increased from south to north, but not an increase in the strict sense. This trend may be related to the impact of management measures such as manual salvage of algae.

Comprehensive water quality assessment based on the modified Nemerow index method

With the modified Nemerow index method, comprehensive water quality assessment was carried out based on the standard limit of Level II water quality in the *Environmental*

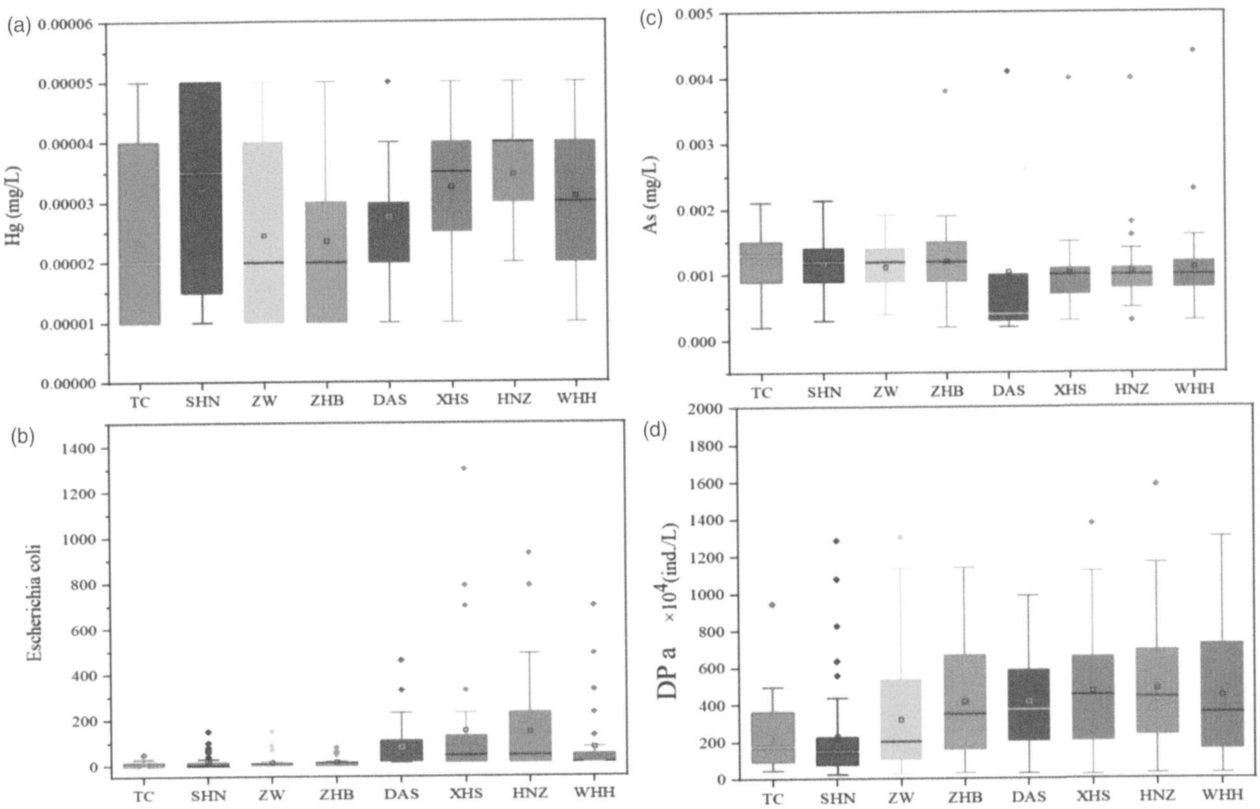

Figure 4 | Average concentration and standard deviation of (a) Hg; (b) As; (c) *Escherichia coli*; and (d) DP-a in 8 water quality monitoring sections of the MR-SNWDP from 2015 to 2018.

Quality Standards for Surface Water (GB3838-2002) and 26 indicators, including 24 selected water quality assessment indicators and two added indicators (comprehensive bio-toxicity and density of planktonic algae). The water quality was divided into five categories (A1, A2, A3, B1 and B2) as per the GB standards, and through value assignment and weighted calculation, the water quality standard limits for different categories of water quality were obtained. These values of water quality standard limits (for rivers) were then used to calculate the modified Nemerow index corresponding to various water quality levels, as shown in Table 2. When the calculation result of the modified Nemerow index is less than 1, the water quality is considered as A1; that is, the best water quality; when the calculation result of the modified Nemerow index is between 1 and 2, the water quality is categorized into A2, which means the water is 'safe and sensitive', and so on.

The overall assessment results of the water quality of the main canal in 2018 were shown in Figure 5. The overall *P*-value was between 0.5 and 2.0, suggesting a safe and good water quality. Different sections presented certain changes and differences in different seasons, and the Nemerow index scores of TC, SHN, DAS and WHH experienced obvious seasonal changes and underwent a typical increase in spring and autumn. The water quality was ranked by season according to the scores: winter < summer < spring < autumn. There are two possible causes: first, the impact of climate, light and temperature; second, the multidimensional

Table 2 | The modified Nemerow index and classification of water quality categories

Water quality category	A1 Very good	A2 Safe and sensitive	A3 Safe but fragile	B1 Poor	B2 Badly
Nemerow index P	$P \leq 1.0$	$1.0 < P \leq 2.0$	$2.0 < P \leq 3.0$	$3.0 < P \leq 4.0$	$4.0 < P \leq 5.0$

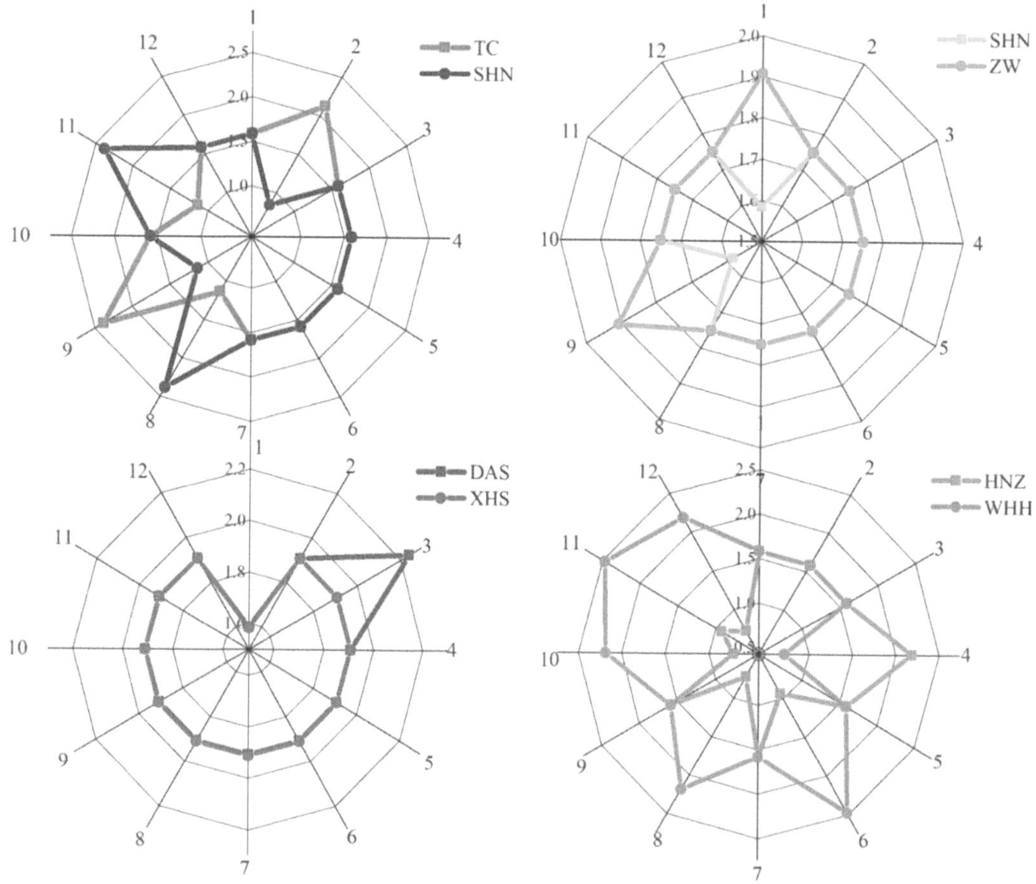

Figure 5 | Results of applying the modified Nemerow index method to assess the 8 monitoring stations of the main canal of the MR-SNWDP from January to December 2018.

comprehensive effect of water quality factors. In terms of natural environmental conditions, water quality indicators such as water temperature, dissolved oxygen, permanganate index, ammonia nitrogen and density of planktonic algae had great influences on the comprehensive calculation result of the water quality index (Nong *et al.* 2020); then in terms of project operation and management, uncertain factors such as the control over risk sources and mechanical operation might also affect the water quality (Kuo *et al.* 2019). As a small-scale ecological system, the open canal creates a suitable growing environment for aquatic plants in spring and autumn when the light and temperature are comfortable. Nonetheless, the ecological system is still unstable at the beginning of the project operation, and there might be an explosive growth of aquatic animals and plants without a natural enemy or artificial intervention. Using the modified Nemerow index method to analyze and assess the water quality of the main canal can, under the condition that the overall water quality is good, help discover the hidden dangers and pollution risks, in order for the management department to carry out emergency prevention and control or disposal accordingly (Nong *et al.* 2020).

Impact of weight on the assessment results

In routine operation, the MR Administration Bureau used the single-factor assessment method based on the national standards for surface water assessment to select the worst water quality level as the basis for assessment (Figure 6(a)), but this method might fail to reach an accurate assessment of the overall water quality. The assessment result obtained by this method showed that the water quality of the monitoring stations was at a poor level, meeting the GB standards for Class IV or Class V water, as shown in Figure 6(a).

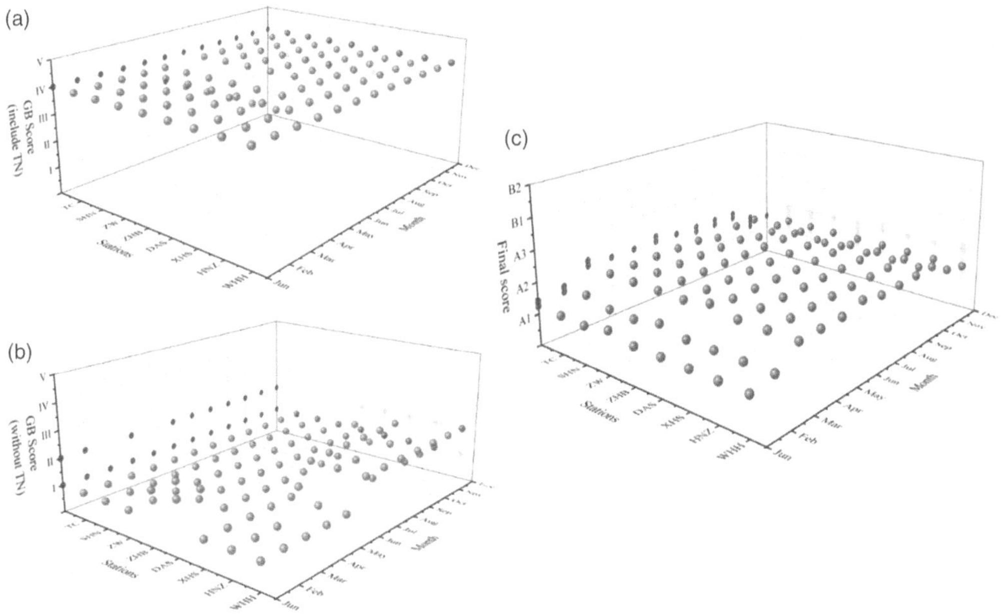

Figure 6 | Results of the comparison of the traditional single-factor method with (a) or without (b) TN index and the results with the Nemerow index method (c) with the 26 indicators from Table 1.

This result, however, was subject to the impact of the high value of TN. With the impact of TN excluded, the water quality improved considerably and met the GB standards for Level I and Level II water, as shown in Figure 6(b). The assessment result obtained by the modified Nemerow method was shown in Figure 6(c), where A1, A2, and A3 represent 'very good', 'safe and sensitive', and 'safe but fragile' water quality, respectively; B1 and B2 represent 'poor' and' bad' water quality, respectively. Comparison of these three figures shows that the assessment result obtained by the modified Nemerow method indicates the poor water quality measured in the stations from March to October, which more accurately reflects the overall water quality and ecological conditions in these stations than the single-factor method and excluded the impact of the abnormally high indicator of TN concentration. Moreover, the proposed method can reflect the dynamic changes of the water quality along the open canal of the Middle Route. Different from natural water delivery systems, the open canal is an artificial system that has been in service for merely five years, so its long-term ecological impact should be given attention. The comparison verified the modified Nemerow method and confirmed its validity to guide real-world engineering practice.

CONCLUSION

In this paper, the water quality of the main canal of the MR-SNWDP from its start of service to March 2019 was analyzed based on 26 water quality indicators in eight key stations along the canal. The modified Nemerow index method was employed to provide comprehensive water quality assessment, and the major conclusions are as follows.

(1) During the study period, the water quality of the MR-SNWDP has stayed at 'Level II' or 'Level I' water quality level; that is, a 'good' or 'excellent' level, of the GB Water Quality Standards (with the impact of TN concentration excluded); physicochemical and heavy metal indicators all met the GB standards of Level II water quality; water safety was assured during the monitoring and assessment period.
(2) Compared with the single-factor assessment method, the modified Nemerow index method delivers a more objective assessment result and better reflects the changes of water quality in spring and autumn; it also demonstrates the contribution of indicators like the heavy metal content and nutrient salt content to the overall water

quality, thereby providing a more comprehensive view of the impacts of different factors on the water quality than the single-factor method. Despite the different results these two methods reached, both showed that the water quality of the Middle Route remained at a good or excellent level since it has been put into service.

(3) Major water quality indicators that deserve special attention were defined, including temperature, permanganate index, total phosphorus, total nitrogen, mercury, arsenic and density of planktonic algae. Moreover, it is advisable that administrations pay more attention to environmental indicators like the temperature and TN/TP that might cause abnormal reproduction of algae and shellfish.

ACKNOWLEDGEMENTS

The study was supported by the IWHR Research & Development Support Program (JZ0145B772017). The writers also acknowledge the assistance of anonymous reviewers.

DATA AVAILABILITY STATEMENT

All relevant data are included in the paper or its Supplementary Information.

REFERENCES

Angelakis, A. N., Do Monte, M. H. F. M., Bontoux, L. & Asano, T. 1999 The status of wastewater reuse practice in the Mediterranean basin: need for guidelines. *Water Research* **33**, 2201–2217.

Cheng, B. F., Xia, R., Zhang, Y., Yang, Z. W., Hu, S., Guo, F. & Ma, S. Q. 2019 Characterization and causes analysis for algae blooms in large river system. *Sustainable Cities and Society* **51**,101707. https://doi.org/10.1016/j.scs.2019.101707.

Kong, M. Y., Zhong, H. P., Wu, Y. X., Liu, G. D., Xu, Y. & Wang, G. X. 2019 Developing and validating intrinsic groundwater vulnerability maps in regions with limited data: a case study from Datong City in China using DRASTIC and Nemerow pollution indices. *Environmental Earth Sciences* **78**.

Kuo, Y. M., Liu, W. W., Zhao, E. M., Li, R. & Munoz-Carpena, R. 2019 Water quality variability in the middle and down streams of Han River under the influence of the Middle Route of South-North Water diversion project, China. *Journal of Hydrology* **569**, 218–229.

Lin, M. L., Lek, S., Ren, P., Li, S. H., Li, W., Du, X., Guo, C. B., Gozlan, R. E. & Li, Z. J. 2017 Predicting impacts of south-to-north water transfer project on fish assemblages in Hongze Lake, China. *Journal of Applied Ichthyology* **33**, 395–402.

Liu, T. Y., Yang, Y. L., Wang, Z. L. & Sun, Y. Q. 2016 Remediation of arsenic(III) from aqueous solutions using improved nanoscale zero-valent iron on pumice. *Chemical Engineering Journal* **288**, 739–744.

Liu, R. Z., Zhang, K., Zhang, Z. J. & Borthwick, A. G. L. 2018 Watershed-scale environmental risk assessment of accidental water pollution: the case of Laoguan River, China. *Journal of Environmental Informatics* **31** (2), 87–96.

Long, Y., Yang, Y. L., Lei, X. H., Tian, Y. & Li, Y. M. 2019 Integrated assessment method of emergency plan for sudden water pollution accidents based on improved TOPSIS, Shannon entropy and a coordinated development degree model. *Sustainability* **11** (2), 510.

Nong, X., Shao, D., Zhong, H. & Liang, J. 2020 Evaluation of water quality in the South-to-North Water Diversion Project of China using the water quality index (WQI) method. *Water Research* **178**, 115781.

Shao, D. G., Yang, H. D., Xiao, Y. & Liu, B. Y. 2014 Water quality model parameter identification of an open channel in a long distance water transfer project based on finite difference, difference evolution and Monte Carlo. *Water Science and Technology* **69**, 587–594.

Shi, B., Jiang, J., Liu, R., Khan, A. U. & Wang, P. 2017 Engineering risk assessment for emergency disposal projects of sudden water pollution incidents. *Environmental Science and Pollution Research International* **24**, 14819–14833.

Tan, X., Xia, X. L., Li, S. Y. & Zhang, Q. F. 2015 Water quality characteristics and integrated assessment based on multistep correlation analysis in the Danjiangkou Reservoir, China. *Journal of Environmental Informatics* **25**, 60–70.

Tang, C. H., Yi, Y. J., Yang, Z. F. & Cheng, X. 2014 Water pollution risk simulation and prediction in the main canal of the South-to-North Water Transfer Project. *Journal of Hydrology* **519**, 2111–2120.

Veettil, A. V. & Mishra, A. K. 2016 Water security assessment using blue and green water footprint concepts. *Journal of Hydrology* **542**, 589–602.

Wang, X., Sugumaran, V., Zhang, H. & Xu, Z. 2017 A capability assessment model for emergency management organizations. *Information Systems Frontiers* **20**, 653–667.

Wei, S. K., Yang, H., Abbaspour, K., Mousavi, J. & Gnauck, A. 2010 Game theory based models to analyze water conflicts in the Middle Route of the South-to-North Water Transfer Project in China. *Water Research* **44**, 2499–2516.

Wu, J. H., Xue, C. Y., Tian, R. & Wang, S. 2017 Lake water quality assessment: a case study of Shahu Lake in the semiarid loess area of northwest China. *Environmental Earth Sciences* **76**, 232. https://doi.org/10.1007/s12665-017-6516-x.

Yang, Y., Lei, X., Long, Y., Tian, Y., Zhang, Y., Yao, Y., Hou, X., Shi, M., Wang, P., Zhang, C., Wang, H. & Quan, J. 2020 A novel comprehensive risk assessment method for sudden water accidents in the Middle Route of the South–North Water Transfer Project (China). *Science of the Total Environment* **698**, 134167.

Zhao, H. G., Huang, Y. C., You, S. C., Wu, Y. F. & Zheng, F. X. 2019 A framework for assessing the effects of afforestation and south-to-north water transfer on nitrogen and phosphorus uptake by plants in a critical riparian zone. *Science of the Total Environment* **651**, 942–952.

Zhu, J., Lei, X. H., Quan, J. & Yue, X. 2019 Algae growth distribution and key prevention and control positions for the Middle Route of the South-to-North Water Diversion Project. *Water* **11**, 1851. https://dx.doi.org/10.3390/w11091851.

First received 11 August 2020; accepted in revised form 17 December 2020. Available online 11 January 2021

Formation of iodinated products in Fe (II)/peroxydisulfate (PDS) system

Zijun Dong, Mu Li, Meifen Chen, Yurong Gu, Chengchun Jiang and Feiyun Sun

ABSTRACT

The transformation of iodide (I^-) and hypoiodous acid (IO^-) by Fe (II)-activated peroxydisulfate (PDS) in the presence of natural organic matter (NOM) was investigated. In the water of a reservoir in Shenzhen, four kinds of organic matter were detected: iodoform > diiodo-bromomethane > dibromo-iodomethane > dichloromethane. As pH increased from 3 to 10, I^- conversion rate decreased from 31.0% to 11.0%. There is basically no iodic acid (IO_3^-) generated under alkaline conditions, probably due to these organic matters. When PDS concentration is low, PDS can oxidize part of I^- to IO^-, and the oxidation of I^- into IO_3^- reaction and continued oxidation reaction exist at the same time, so some IO_3^- is generated in the system. As the proportion of Fe (II) increases, the generation rate of IO_3^- gradually increases. When Fe^{2+}/PDS is 1:5, 2:5, and 3:5, the conversion rates of IO_3^- are 4.7%, 6.5%, and 8.4%, respectively. There are two main reasons for this: (i) the reaction of IO^- produced in the system with humic acid (HA) inhibits the further conversion of IO^- to IO_3^-; (ii) both HA and I- in the system can react with PDS, resulting in a reduction in the oxidation rate of I^-.

Key words | Fe (II), iodide, natural organic matter (NOM), peroxydisulfate (PDS)

Zijun Dong
Meifen Chen
Yurong Gu (corresponding author)
Chengchun Jiang
School of Civil and Environmental Engineering,
Shenzhen Polytechnic,
Shenzhen 518055,
China
E-mail: *gyr0809405018@szpt.edu.cn*

Mu Li
Shenzhen Environmental Science and New Energy Laboratory, Tsinghua-Berkeley Shenzhen Institute,
Tsinghua University,
Shenzhen,
China

Feiyun Sun
School of Civil and Environmental Engineering,
Harbin Institute of Technology (Shenzhen),
Shenzhen 518055,
China

HIGHLIGHTS

- Formation of I-DBPs by Fe (II)/PDS process in real water was evaluated.
- I-DBPs generation in various reaction conditions was examined.
- Existence of HA showed apparent inhibition on IO_3^- generation.

doi: 10.2166/ws.2020.254

GRAPHICAL ABSTRACT

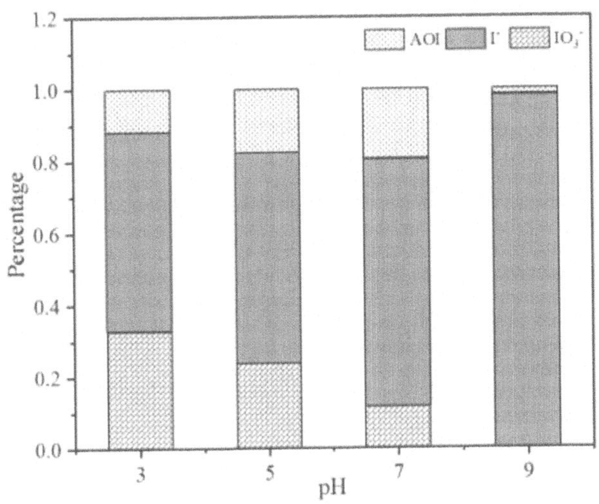

INTRODUCTION

In the natural water environment, the main forms of iodide are iodide ions (I^-) and iodic acid (IO_3^-) (Bichsel & Von Gunten 1999; Richardson *et al.* 2008; Zhao *et al.* 2016). In addition to iodine in natural water, iodine enters water in industrial and medical industries, such as iodinated X-ray contrast media (ICM compounds), causing serious iodine pollution (Pérez & Barceló 2007; Duirk *et al.* 2011). Iodide readily reacts with oxidants to form hypoiodous acid (HOI) such as permanganate ($KMnO_4$), ozone (O_3), UV (Magi *et al.* 1997; Criquet *et al.* 2012; De la Cruz *et al.* 2012; Yang *et al.* 2014). During oxidation, two pathways are conducted: (i) HOI reacts with natural organic matter (NOM) in water to produce a series of iodine by-products (I-DBPs), including IO_3^- and iodomethane, (ii) HOI is further oxidized to iodate (IO_3^-) (Li *et al.* 2017; Dong *et al.* 2019). During the past decade, it has been investigated that I-DBPs are more toxic than chlorinated and brominated by-products, so there is an urgent need for a way to control iodine by-products in water bodies (Smith *et al.* 2010; Gan *et al.* 2015; Liu *et al.* 2017; Hu *et al.* 2018; Huang *et al.* 2019). The generation factors of I-DBPs are affected by the quality of the raw water, the type of oxidant, the pH of the water and the concentration of I− in the water (Tian *et al.* 2014; Dong *et al.* 2018; Hu *et al.* 2019; Wu *et al.* 2019; Zhao *et al.* 2019). The NOM in the water body reacts with the oxidant to produce by-product precursors. The properties of these precursors have a certain influence on the production of iodine by-products (Dong *et al.* 2019). The types of precursor organic matter directly affect the types of by-products. The oxidant can sequentially oxidize the I^- to IO^- and IO_3^- (Li *et al.* 2017). The IO_3^- will not react with organic matter to form iodine by-products, and the reaction rates of hypoiodine with different oxidants are different. Previous studies have investigated that peroxymonosulfate was able to oxidize I^-, the second-order rate constant of reaction was $1.01 \times 10^3\ M^{-1}\ s^{-1}$, while that for HOI was $1.08 \times 10^2\ M^{-1}\ s^{-1}$ (Li *et al.* 2017). Studies have shown that the concentration of iodized trihalomethanes increased with the concentration of I^- in raw water (Hua *et al.* 2006; Allard *et al.* 2016; Kralchevska *et al.* 2016; Liu *et al.* 2017).

Among all oxidants, $SO_4^{\cdot -}$ is a highly active oxidant with unsaturated electronic structure and unstable chemical properties, which has the characteristics of strong oxidation and short half-life. Under neutral conditions, the oxidation–reduction potential of $SO_4^{\cdot -}$ is higher than $\cdot OH$, and most organic pollutants can be completely degraded by $SO_4^{\cdot -}$ (Anipsitakis & Dionysiou 2003). The transformation of pollutants by $SO_4^{\cdot -}$ usually follows three pathways: (i) $SO_4^{\cdot -}$

degrades saturated organics such as alkanes, alcohols, and lipids through hydrogen atom extraction; (ii) $SO_4^{\cdot-}$ passes through unsaturated bonds of unsaturated olefins by addition reaction (House 1962; Rastogi *et al.* 2009; Wu *et al.* 2012; Ding *et al.* 2013; Xie *et al.* 2015); (iii) electron transfer reaction. However, there is little information about I^- conversion in the $SO_4^{\cdot-}$ system oxidation process. The result of rapid conversion of I^- to HOI by $SO_4^{\cdot-}$ indicates that the potential genotoxicity and cytotoxicity of I-DBP are usually several to hundreds of times higher than its chlorinated and brominated analogs. Peroxydisulfate (PDS), as an oxidant that is relatively stable at room temperature, has strong oxidizing properties and high solubility, and it has gradually been used in the environmental field (House 1962; Wu *et al.* 2012). The advantages enable PDS to be used in new types of advanced oxidation technologies. Thermal activation, ultraviolet activation, metal ion and metal oxide activation methods are the most common activation methods reported in the literature. Transitional metal can activate PDS to produce $SO_4^{\cdot-}$ with the activation energy required for metal ion catalysis being low, about 50 kJ/mol (Zhao *et al.* 2019). However, in the Fe^{2+}/PDS system, due to the complexity of NOMs in the water, the types of I-DBPs generated are also more complex, transforming with the variation of environment (Kamarehie *et al.* 2017; Mohagheghian *et al.* 2017). The objectives of this study were (i) to investigate the varieties of I-DBPs in a natural water environment and (ii) to explore the generation of I-DBPs in the Fe (II)/PDS oxidation process under different factors.

MATERIALS AND METHODS

Chemicals

Sodium persulfate (PDS) with purity 99%, NaClO (6%–14% active chlorine), KI with purity >99% and KIO_3 with purity 99.8% were purchased from Shanghai Aladdin. Ferrous sulfate with purity >99%, ferric sulfate with purity 97%, nitrobenzene with purity >99%, sodium hydroxide with purity >96%, and sulfuric acid (purity 95%–98%) were purchased from Sinopharm Chemical Reagent Co. Ltd. Methanol, acetonitrile and acetic acid with purity >99% were purchased from Merck. All solutions were prepared using deionized water with 18.2 MΩ/cm Milli-Q water.

Formation of iodinated products

All the experiments were conducted in a 250 mL batch reactor at temperature 25 °C. Add a pre-configured buffer solution containing 4 mM sodium tetraborate/potassium dihydrogen phosphate to a 250 mL iodine flask, add a calculated volume of KI standard solution, and prepare the concentration of KI required for the experiment. Use 0.1 mol/L H_2SO_4 or 0.1 mol/L NaOH to adjust the pH of the prepared KI solution to the set value. The sampling amount is determined according to the amount of target substance required to determine the index, and a certain volume of sodium sulfite needs to be added immediately after sampling to terminate the reaction. After the end, the pH change is less than 0.2.

When measuring IO_3^-, add a calculated volume of Fe (II) to a 250 mL iodine flask, then add a calculated volume of PDS, and at the same time start timing sampling, take 1 mL and add it to the liquid phase containing 2,6-dichlorophenol and anhydrous sodium sulfite. In a vial, measure hypoiodic acid, take 5 mL and add it to an ion chromatography tube containing anhydrous sodium sulfite, and measure iodide ion and hypoiodic acid. In order to ensure the reliability of the experimental results, each group of experiments was repeated three times, and the error of the experimental results was controlled within 5%, and a blank experiment was set.

Analysis

I^-, IO^- and IO_3^-

I^-, IO_3^- and IO^- were measured by ion chromatography (ICS1500). The specific chromatographic conditions are a Dionex ICS1500 ion chromatograph (USA); isocratic leaching is used, the eluent is 9 mmol/L Na_2CO_3; the flow rate of the eluent is 1 mL/min; the suppression method is conductivity suppression, which is caused by electrolytic water producing H^+, the suppressor is an ASRS-Ultra 2 mm external water anion suppressor; column temperature: 35 °C; the detector is a conductivity detector; detector temperature is

35 °C; chromatographic column: IonpacAS9 type protection column (4 mm × 4 mm); IonPacAS9-HC type separation column (250 mm × 4 mm); injection volume is 1,000 μg/L; method detection limit is 0.3 μg/L. The detection of IO^- used the method of reacting IO^- with 2,6-dichlorophenol to produce 4-iodo-2,6-dichlorophenol, by measuring the amount of 4-iodo-2,6-dichlorophenol. The 4-iodo-2,6-dichlorophenol was measured by Waters high performance liquid chromatography under the conditions of: C18 column (4.6 mm × 250 mm × 5 μm), and the mobile phase is V(acetonitrile): V(water) = 7:3, ultraviolet detection wavelength is 212 nm, flow rate is 1 mL/min, injection volume is 100 uL.

Formation of I-DBPs in water

The determination of I-DBPs is carried out by combined use of gas and quality. Gas chromatographic conditions: HP-5MS capillary column (30.0 m × 0.25 mm × 0.25 μm), electron capture detector, detector temperature is 300 °C, inlet temperature is 200 °C, program temperature is 40 °C, hold for 5 min, to raise 1 °C/min to 45 °C, hold for 3 min, raise 10 °C/min to 135 °C, then raise 25 °C/min to 220 °C, inject 1 μL, splitless injection. Mass spectrometry conditions: EI ion source, ion source temperature 225 °C, quadrupole temperature 150 °C, chromatography–mass spectrometer connection temperature 170 °C, electron energy 70 eV, solvent delay 3 min, scan mass range m/z 20–500, scan mode select ion monitoring (m/z).

Extraction conditions by gas mass spectrometry: MTBE is used as the extractant. At the end of the reaction, a 30 mL sample and a bottle with a Teflon cap are taken, 6 g of anhydrous sodium carbonate is added to shake, and a 3 mL pure MTBE solution is added to the sample with a pipette gun. In the test, shake for 20 minutes in a vortex shaker and stand still for 20 minutes. Take 1 mL of the upper layer organic solution in the sample bottle with the air quality for detection.

AOI

Absorbable organic halogen (AOI) was measured with a Multi X 2500 total organic halogen analyzer (Analytic Jena, Germany). An APU28 carbon column was used for sample adsorption. Add nitric acid stock solution to the sample, the pH value of the whole sample is less than 2, take a 50 mL sample in the pressure filter, and set the program. After the sample passes through the APU28, the AOI in it is completely adsorbed on the APU28 carbon column. Then elute the ions in the carbon column with 25 mL sodium nitrate stock solution, take out the sample carbon column at the end of the desorption, and absorb the surface moisture on the filter paper; put the sample carbon column into the combustion tube, mineralize at 950 °C, AOI is converted into HI gas; HI gas is dehydrated by concentrated sulfuric acid, and detected by high-sensitivity microcoulometric titration. In order to ensure reliable measurement results, 2-iodophenol is used as the standard substance for calibration.

RESULTS AND DISCUSSION

Formation of I-DBPs in real water

The formation of I-DBPs is closely related to the type of organic matter in the water. In the actual water body, the organic matter contained in it is relatively complex, including multiple functional groups: benzene rings, amino groups, carboxyl groups, etc. The type of organic matter may be different from the products formed by PDS oxidation, but the possible paths and mechanisms for functional groups are similar. To investigate I-DBPs formation during oxidation of PDS by Fe (II) in real water, I^- was spiked into the river water from a reservoir from Shenzhen. Six typical iodinated organic compounds (bromochloroiodomethane, monobromodiiodomethane, dibromomonoiodomethane, monochlorodiiodomethane, dichloromonoiodomethane, iodoform) were monitored.

Figure 1 shows the determination of I-DBPs at an initial PDS concentration 100 μM, Fe (II) concentration 80 μM and pH 3, over 120 min. The results suggest that the concentration order of the detected four kinds of organic matter is: iodoform > diiodo-bromomethane > dibromo-iodomethane > dichloromethane, the maximum concentration of iodoform is 0.49 μM, and the concentration of iodoform can induce cell mutagenesis and endanger human health. In addition to the detected iodine by-products, there are also iodine by-products including

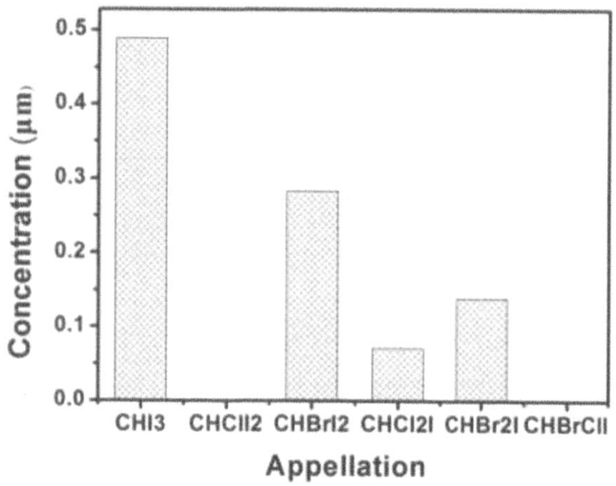

Figure 1 | Determination of iodomethane in actual water (pH = 3, [Fe (II)] = 80 μM, T = 25 °C, t = 120 min).

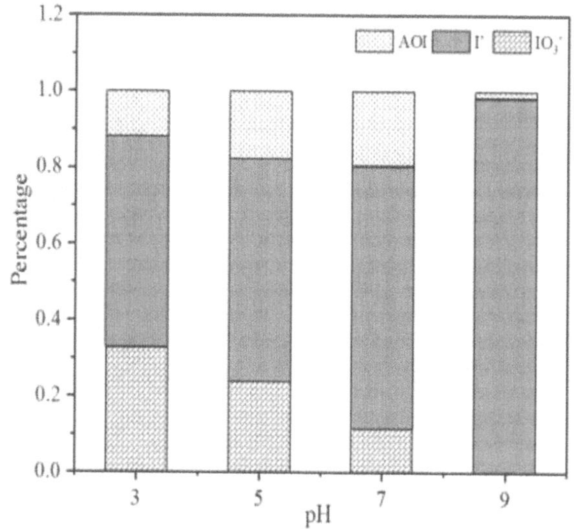

Figure 2 | Effect of pH on the formation of I^- transformation ([PDS] = 100 μM, [I^-] = 10 μM, [Fe^{2+}/PDS] = 4:5, [HA] = 5 mg/L, T = 25 °C).

diiodine-bromomethane, iodine dichloromethane, and iodine dibromomethane in the water, all of which are highly toxic. Due to the instability of the actual water body, all the iodine-containing organic substances are characterized by an index of the sum of iodine by-products in the experiment – Adsorbable Organic Iodine (AOI).

Effect of initial pH

The oxidation effect of PDS at different pH is different. At the same time, pH also affects the type of active iodine produced by PDS oxidizing iodine ions, and the type of active iodine is greatly different in the degree of difficulty of being reacted by the oxidant. The pH affects the ability of PDS to oxidize I^-, the presence of Fe (II) in the experimental system, and the rate of PDS oxidation of active iodine. Therefore, the types and amounts of I-DBPs produced under different pH conditions will be quite different.

The change of formation of I^- transformation with pH in the Fe (II)/PDS system is illustrated in Figure 2. The ratio of the Fe (II) transformation to PDS was 4:5 as pH was increased from 3 to 10. As the pH increases, the total amount of AOI gradually increases and then slowly decreases, IO_3^- gradually decreases as the pH increases, I^- gradually increase as the pH increases and the concentration of IO^- is barely detectable in the solution. For IO_3^-, 31% of the I^- are converted to IO_3^- at pH 3. While in organic-free systems, the IO_3^- generation rate is 56% and gradually decreases at pH = 4, 5, 6, and 7. Especially, there is basically no IO_3^- generated under alkaline conditions, probably due to the organic matter. When the system generates IO_3^-, it is quickly captured and cannot be oxidized. For I^-, the conversion efficiency of I^- gradually decreases with increasing pH. As the pH increases from 3 to 9, the remaining amount of I^- rises from 57% to 98%. Compared with the absence of organic matter, the conversion efficiency of I^- is also greatly reduced at the same pH. This is because the organic substances in the system react with sulfate radicals generated by Fe (II)/PDS, and the free radicals participating in the oxidation reaction of I^- decrease, so that the conversion rate of I^- decreases. No IO^- was detected in the system with organic matter. This is because a large amount of organic matter causes a part of the IO^- produced by the reaction to be captured and a part of it to be oxidized to IO_3^-. It can be seen from the iodine conservation curve that the total iodine amount of I^- plus IO_3^- plus AOI is close to the total amount of iodine, basically reaching iodine conservation.

Effect of initial PDS dosage

Figure 3 shows the effect of PDS dosage on the I^- transformation in the Fe (II)/PDS system, at a constant Fe (II) dosage of 20 μM, and I^- initial concentration of 10 μM. The ratio of the Fe (II) transformation to PDS was 4:5 at pH 3. When the concentration of the PDS is low, PDS can oxidize part of I^- to IO^-, and the oxidation of I^- into IO_3^- reaction and the continued oxidation reaction exist at the same time, so some IO_3^- is generated in the system. However, due to the limited concentration of PDS, IO^- that has not been oxidized to IO_3^- can quickly react with organic matter to form I-DBPs. The concentration of IO_3^- increases with the increase of the PDS concentration, and it stabilizes at 100 μM, indicating that the higher the PDS concentration, the more thorough the oxidation reaction that occurs, and the higher the I conversion rate, the saturation reaches 100 μM. For I^-, the higher the PDS concentration, the lower the iodine ion concentration, indicating that more I^- are oxidized. The sum of I^- and IO_3^- concentration after 40 μM is almost unchanged, indicating that the PDS concentration has a greater influence on the first step of the iodine oxidation reaction, and has little effect on the conversion of IO^- to I-DBPs. The law of AOI decreases with the increase of oxidant concentration, which is the largest at 60, indicating that under this condition, secondary iodine easily reacts with organic matter to form iodine by-products.

Effect of initial Fe (II)/PDS molar ratio

Figure 4 shows the effect of Fe (II)/PDS on I^- transformation in the Fe (II)/PDS system, at a constant PDS dosage of 100 μM, I^- initial concentration of 10 μM at pH 3. As the proportion of Fe (II) increases, the generation rate of IO_3^- gradually increases. When Fe^{2+}/PDS is 1:5, 2:5, and 3:5, the conversion rates of IO_3^- are 4.7%, 6.5%, and 8.4, respectively. When Fe^{2+}/PDS reaches 5:5, about 33% of I^- are oxidized to IO_3^-. When Fe^{2+}/PDS reaches 6:5, 36% of the I^- are converted to IO_3^-, indicating that at Fe^{2+}/PDS ratio greater than 4:5, a large amount of I^- are converted into IO^- in the experimental system, and then further converted into IO_3^-, and when Fe^{2+}/PDS is small, the activation generates fewer sulfate radicals and the ability to oxidize I^- is insufficient, so the generation rate of IO_3^- is small.

The conversion rate of I^- is similar to that of IO_3^-. It gradually increases with the increase of Fe^{2+}/PDS, and then tends to be stable. It can be seen from the figure that when Fe^{2+}/PDS is 6:5, the residual amount of iodide ion is the smallest at 48%, and the conversion rate is the largest. When the Fe^{2+}/PDS ratio is 1:5, the residual amount of iodide ion reaches 73%.

The AOI in the experimental system gradually increases with the increase of the proportion of ferrous iron and then decreases. When the Fe^{2+}/PDS ratio is 3:5, the AOI generation rate is 30% at the maximum, and the Fe^{2+}/PDS ratio

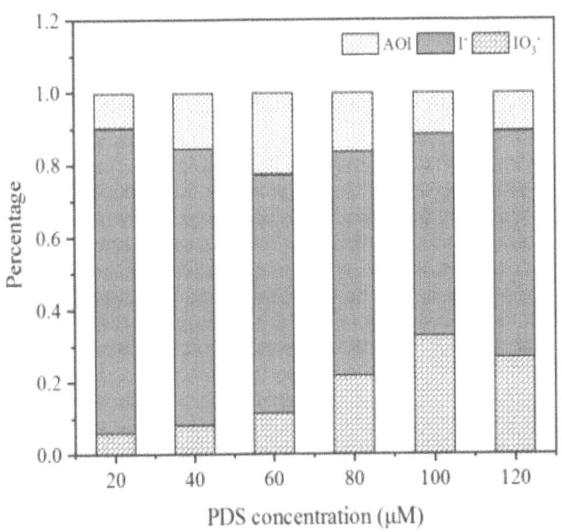

Figure 3 | Effect of PDS dosage on the formation of I^- transformation (pH 3, $[I^-] = 10$ μM, $[Fe^{2+}/PDS] = 4{:}5$, [HA] = 5 mg/L, $T = 25$ °C).

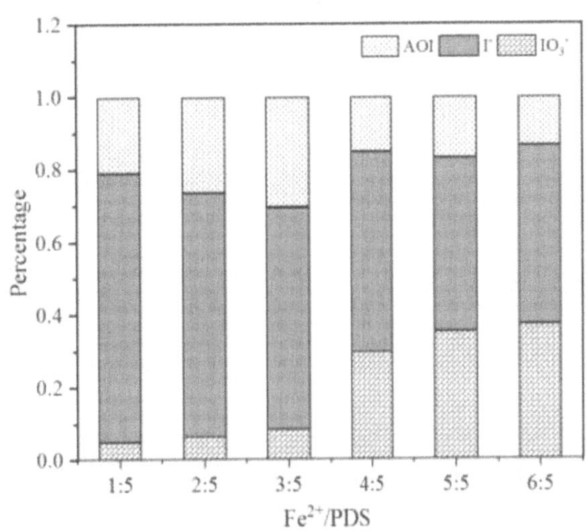

Figure 4 | Effect of Fe (II)/PDS on the formation of I^- transformation (pH 3, [PDS] = 100 μM, $[I^-] = 10$ μM, [HA] = 5 mg/L, $T = 25$ °C).

is 4:5, 5:5 and 6:5. The generation rates of AOI are 15.9%, 15.7% and 14.0%.

Effect of NOM

The effect of NOM on I^- transformation in the Fe(II)/PDS system was examined in the presence of commercial humic acid (HA; used as a representative NOM), at pH 3. As can be seen in Figure 5, as the concentration of HA increases, the AOI in the system gradually increases. When the HA concentration reaches 1 mg/L, the AOI reaches the maximum. When the HA concentration in the experimental system is increased again, AOI concentration remains basically unchanged. The amount of IO_3^- produced gradually decreases with increasing HA concentration. There are two main reasons for this: (i) the reaction of IO^- produced in the system with HA inhibits the further conversion of IO^- to IO_3^-; (ii) both HA and I^- in the system can react with PDS, resulting in a reduction in the oxidation rate of I^-.

From this we can conclude that the total organic carbon of the water body affects the generation of I-DBPs in the water. It is not that the higher the TOC, the easier it is for the water body to produce I-DBPs. We can also control I-DBPs by controlling the total organic carbon product formation.

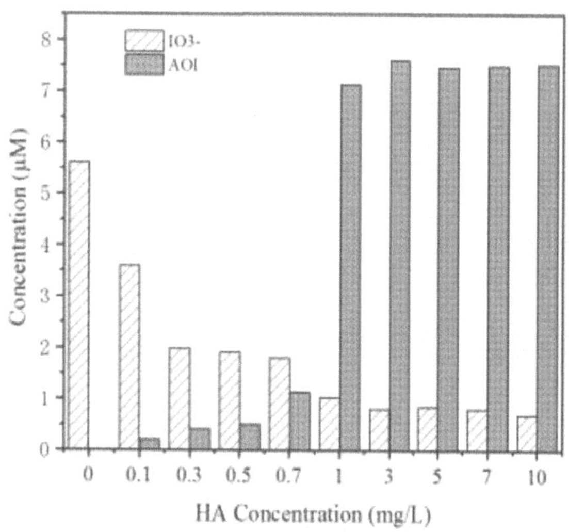

Figure 5 | Effect of HA on the formation of I^- transformation (pH 3, [PDS] = 100 μM, [I^-] = 10 μM, [Fe^{2+}/PDS] = 4:5, T = 25 °C).

CONCLUSIONS

This study investigated the Fe (II) activated PDS process under different pH, Fe (II)/PDS, PDS dosage and NOM conditions. Based on the experimental results obtained, the following conclusions can be obtained:

(1) The experiment results verified the generation of AOI in the case of adding NOM to the Fe (II)/PDS oxidation system. It was found that under acidic conditions, as the AOI formation rate is larger, I^- production is more.
(2) Under alkaline conditions, although the AOI generation rate is small, the conversion rate of iodide ions is small. When the concentration of oxidant in the experimental system is increased, the AOI generation amount gradually increases and then decreases.
(3) The AOI generation increases with the proportion of Fe (II). When the ratio is 3:5, the generation rate of AOI is at most 30%. The generation of AOI gradually increases with the increase of HA concentration, and when the concentration is 1 mg/L, the AOI concentration is the largest.

ACKNOWLEDGEMENT

We thank funding support from Guangdong Basic and Applied Basic Research Foundation (Grant No. 2020A1515011545), Shenzhen Water Group Program (Grant No. 2019-131-F) and Shenzhen Polytechnic Program (Grant No. 6019310013K0).

DATA AVAILABILITY STATEMENT

All relevant data are included in the paper or its Supplementary Information.

REFERENCES

Allard, S., Criquet, J., Prunier, A., Falantin, C., Le Person, A., Tang, J. Y.-M. & Croué, J.-P. 2016 Photodecomposition of iodinated contrast media and subsequent formation of toxic iodinated moieties during final disinfection with chlorinated oxidants. *Water Research* **103**, 453–461.

Anipsitakis, G. P. & Dionysiou, D. D. 2003 Degradation of organic contaminants in water with sulfate radicals generated by the conjunction of peroxymonosulfate with cobalt. *Environmental Science & Technology* **37** (20), 4790–4797.

Bichsel, Y. & von Gunten, U. 1999 Oxidation of iodide and hypoiodous acid in the disinfection of natural waters. *Environmental Science & Technology* **33** (22), 4040–4045.

Criquet, J., Allard, S., Salhi, E., Joll, C. A., Heitz, A. & von Gunten, U. 2012 Iodate and iodo-trihalomethane formation during chlorination of iodide-containing waters: role of bromide. *Environmental Science & Technology* **46** (13), 7350–7357.

De la Cruz, N., Giménez, J., Esplugas, S., Grandjean, D., de Alencastro, L. F. & Pulgarín, C. 2012 Degradation of 32 emergent contaminants by UV and neutral photo-fenton in domestic wastewater effluent previously treated by activated sludge. *Water Research* **46** (6), 1947–1957.

Ding, Y., Zhu, L., Wang, N. & Tang, H. 2013 Sulfate radicals induced degradation of tetrabromobisphenol A with nanoscaled magnetic $CuFe_2O_4$ as a heterogeneous catalyst of peroxymonosulfate. *Applied Catalysis B: Environmental* **129**, 153–162.

Dong, H., Qiang, Z., Liu, S., Li, J., Yu, J. & Qu, J. 2018 Oxidation of iopamidol with ferrate (Fe(VI)): kinetics and formation of toxic iodinated disinfection by-products. *Water Research* **130**, 200–207.

Dong, Z., Jiang, C., Yang, J., Zhang, X., Dai, W. & Cai, P. 2019 Transformation of iodide by Fe(II) activated peroxydisulfate. *Journal of Hazardous Materials* **373**, 519–526.

Duirk, S. E., Lindell, C., Cornelison, C. C., Kormos, J., Ternes, T. A., Attene-Ramos, M., Osiol, J., Wagner, E. D., Plewa, M. J. & Richardson, S. D. 2011 Formation of toxic iodinated disinfection by-products from compounds used in medical imaging. *Environmental Science & Technology* **45** (16), 6845–6854.

Gan, W., Sharma, V. K., Zhang, X., Yang, L. & Yang, X. 2015 Investigation of disinfection byproducts formation in ferrate(VI) pre-oxidation of NOM and its model compounds followed by chlorination. *Journal of Hazardous Materials* **292**, 197–204.

House, D. A. 1962 Kinetics and mechanism of oxidations by peroxydisulfate. *Chemical Reviews* **62** (3), 185–203.

Hu, S., Gong, T., Xian, Q., Wang, J., Ma, J., Li, Z., Yin, J., Zhang, B. & Xu, B. 2018 Formation of iodinated trihalomethanes and haloacetic acids from aromatic iodinated disinfection byproducts during chloramination. *Water Research* **147**, 254–263.

Hu, J., Wang, C., Ye, Z., Dong, H., Li, M., Chen, J. & Qiang, Z. 2019 Degradation of iodinated disinfection byproducts by VUV/UV process based on a mini-fluidic VUV/UV photoreaction system. *Water Research* **158**, 417–423.

Hua, G., Reckhow, D. A. & Kim, J. 2006 Effect of bromide and iodide ions on the formation and speciation of disinfection byproducts during chlorination. *Environmental Science & Technology* **40** (9), 3050–3056.

Huang, Y., Zhang, Y., Zhou, Q., Li, A., Shi, P., Qiu, J. & Pan, Y. 2019 Detection, identification and control of polar iodinated disinfection byproducts in chlor(am)inated secondary wastewater effluents. *Environmental Science: Water Research & Technology* **5** (2), 397–405.

Kamarehie, B., Mohamadian, J., Mousavi, S. A., Asgari, G. & Shahamat, Y. D. 2017 Aniline degradation from aqueous solution using electro/Fe^{2+}/peroxydisulphate process. *Desalination and Water Treatment* **80**, 337–343.

Kralchevska, R. P., Sharma, V. K., Machala, L. & Zboril, R. 2016 Ferrates (Fe^{VI}, Fe^{V}, and Fe^{IV}) oxidation of iodide: formation of triiodide. *Chemosphere* **144**, 1156–1161.

Li, J., Jiang, J., Zhou, Y., Pang, S.-Y., Gao, Y., Jiang, C., Ma, J., Jin, Y., Yang, Y., Liu, G., Wang, L. & Guan, C. 2017 Kinetics of oxidation of iodide (I^-) and hypoiodous acid (HOI) by peroxymonosulfate (PMS) and formation of iodinated products in the PMS/I^-/NOM system. *Environmental Science & Technology Letters* **4** (2), 76–82.

Liu, S., Li, Z., Dong, H., Goodman, B. A. & Qiang, Z. 2017 Formation of iodo-trihalomethanes, iodo-acetic acids, and iodo-acetamides during chloramination of iodide-containing waters: factors influencing formation and reaction pathways. *Journal of Hazardous Materials* **321**, 28–36.

Magi, L., Schweitzer, F., Pallares, C., Cherif, S., Mirabel, P. & George, C. 1997 Investigation of the uptake rate of ozone and methyl hydroperoxide by water surfaces. *The Journal of Physical Chemistry A* **101** (27), 4943–4949.

Mohagheghian, A., Naghipour, D., Moslemzadeh, M. & Taghavi, K. 2017 Photochemical degradation of 2,4-dichlorophenol in aqueous solutions by Fe^{2+}/peroxydisulfate/UV process. *International Journal of Engineering* **30** (1), 15–22.

Pérez, S. & Barceló, D. 2007 Fate and occurrence of X-ray contrast media in the environment. *Analytical and Bioanalytical Chemistry* **387** (4), 1235–1246.

Rastogi, A., Al-Abed, S. R. & Dionysiou, D. D. 2009 Sulfate radical-based ferrous–peroxymonosulfate oxidative system for PCBs degradation in aqueous and sediment systems. *Applied Catalysis B: Environmental* **85** (3–4), 171–179.

Richardson, S. D., Fasano, F., Ellington, J. J., Crumley, F. G., Buettner, K. M., Evans, J. J., Blount, B. C., Silva, L. K., Waite, T. J., Luther, G. W., McKague, A. B., Miltner, R. J., Wagner, E. D. & Plewa, M. J. 2008 Occurrence and mammalian cell toxicity of iodinated disinfection byproducts in drinking water. *Environmental Science & Technology* **42** (22), 8330–8338.

Smith, E. M., Plewa, M. J., Lindell, C. L., Richardson, S. D. & Mitch, W. A. 2010 Comparison of byproduct formation in waters treated with chlorine and iodine: relevance to point-of-use treatment. *Environmental Science & Technology* **44** (22), 8446–8452.

Tian, F.-X., Xu, B., Lin, Y.-L., Hu, C.-Y., Zhang, T.-Y. & Gao, N.-Y. 2014 Photodegradation kinetics of iopamidol by UV irradiation and enhanced formation of iodinated disinfection by-products in sequential oxidation processes. *Water Research* **58**, 198–208.

Wu, J., Zhang, H. & Qiu, J. 2012 Degradation of Acid Orange 7 in aqueous solution by a novel electro/Fe^{2+}/peroxydisulfate process. *Journal of Hazardous Materials* **215–216**, 138–145.

Wu, Y., Zhu, S., Zhang, W., Bu, L. & Zhou, S. 2019 Comparison of diatrizoate degradation by UV/chlorine and UV/chloramine processes: kinetic mechanisms and iodinated disinfection byproducts formation. *Chemical Engineering Journal* **375**, 121972.

Xie, P., Ma, J., Liu, W., Zou, J., Yue, S., Li, X., Wiesner, M. R. & Fang, J. 2015 Removal of 2-MIB and geosmin using UV/persulfate: contributions of hydroxyl and sulfate radicals. *Water Research* **69**, 223–233.

Yang, Y., Komaki, Y., Kimura, S. Y., Hu, H.-Y., Wagner, E. D., Mariñas, B. J. & Plewa, M. J. 2014 Toxic impact of bromide and iodide on drinking water disinfected with chlorine or chloramines. *Environmental Science & Technology* **48** (20), 12362–12369.

Zhao, X., Salhi, E., Liu, H., Ma, J. & von Gunten, U. 2016 Kinetic and mechanistic aspects of the reactions of iodide and hypoiodous acid with permanganate: oxidation and disproportionation. *Environmental Science & Technology* **50** (8), 4358–4365.

Zhao, X., Jiang, J., Pang, S., Guan, C., Li, J., Wang, Z., Ma, J. & Luo, C. 2019 Degradation of iopamidol by three UV-based oxidation processes: kinetics, pathways, and formation of iodinated disinfection byproducts. *Chemosphere* **221**, 270–277.

First received 24 July 2020; accepted in revised form 22 September 2020. Available online 16 October 2020

Impacts of precipitation and topographic conditions on the model simulation in the north of China

Long Sun IWA, Zhijia Li, Ke Zhang and Tingting Jiang

ABSTRACT

The evaluation of hydrological models for a specific catchment is normally based on the model performance according to the selected performance criteria. However, the catchment rainfall-runoff characteristics could be used for the selection of a suitable hydrological model in study area, which, also, for the problem solve of the model application in ungauged basins. In this study, six conceptual models were applied in three semi-humid or semi-arid catchments to investigate the correlation between catchment characteristics and model structure selection. In addition, the impacts of precipitation and topography in model simulation were analyzed. The results show that runoff generation are highly impacted by catchment topographic index and land cover change, and the influence of slope for river channel is greater than mean slope for the whole catchment due to the runoff generation for partial area. For the catchments under similar climate condition, the impact of topographic features for runoff generation process is greater than the difference of precipitation. It indicates that for a specific catchment, the selection of appropriate model should base on better understanding of the rainfall-runoff relationship. The method of incorporating additional runoff generation module in the traditional model can significantly improve the accuracy of flood simulation.

Key words | flood forecasting, infiltration-excess runoff, semi-humid or semi-arid region, topographic characteristics

Long Sun IWA (corresponding author)
Zhijia Li
Ke Zhang
College of Hydrology and Water Resources,
Hohai University, Nanjing,
China
E-mail: *sunlongmwr@163.com*

Long Sun
Information Center (Hydrology Monitor and Forecast Center), Ministry of Water Resources,
Beijing,
China

Tingting Jiang
Hangzhou Municipal Yuhang District Bureau of Forestry and Water Resources,
Hangzhou,
China

HIGHLIGHTS

- Six conceptual models were applied in three semi-humid or semi-arid catchments.
- The model obviously led to better model performance.
- Runoff generation and production process are highly influenced by catchment topographic index.
- The influence of slope for river channel is greater than mean slope.
- The saturation-excess and infiltration-excess runoff mechanism.

INTRODUCTION

Hydrological models with different model structures are widely used around the world. The hydrologists have developed various conceptual hydrological models since the first application of the Stanford model (Singh 1997; Adnan *et al.* 2020). Conceptual hydrological models are effective tools to explore and understand the complex hydrological cycle processes and mechanisms, and they are also effective ways to solve the key problems in hydrology (Kavetski *et al.* 2006; Liu *et al.* 2017). Hydrological processes vary enormously across different landscapes (Winter 2010). For example, two general runoff generation mechanisms, saturation-excess runoff and infiltration-excess runoff, may

doi: 10.2166/ws.2020.284

occur simultaneously in semi-humid or semi-arid regions (Horton 1933; Huang *et al.* 2016). As the rainfall-runoff relationship of a catchment is highly related to the topographic information, the investigation of the influences of topography and climate conditions on model performance could help us to better understand the model and can also provide information for the region with similar climate condition and catchment characteristics.

Catchment geomorphic properties play a significant role in rainfall-runoff simulation, which is one of the most important factors that should be taken into account in the application of hydrological models (Euser *et al.* 2015; Chao *et al.* 2019). Different catchments always have different underlying surface conditions. A great number of studies have been carried out on the measurement of catchment similarity (Hrachowitz *et al.* 2013; Alhassan & Jin 2020). It is assumed that the catchments are similar if the differences for the underlying surface characteristics is within a certain range. Similar catchments have similar rainfall-runoff behavior that can be simulated using the same model and same parameters. However, if the differences are greater than a certain value, the model structures used for different catchments should be modified (Dunne & Black 1970). Chirico *et al.* found that in Maheuangi catchment, a single power-law function is not sufficient to reflect the soil lateral water transport capacity, dual-power law parallel structure can describe the non-linear drainage characteristics (Chirico *et al.* 2003); Clark *et al.* studied the Panola mountain area and their result shows that the non-linear function of runoff can be avoided by using a simple linear reservoir parallel structure (Clark *et al.* 2009). In order to investigate the relationship between catchment characteristics and conceptual model structures, Fenicia *et al.* used 12 model structures, which including single reservoir structure, tandem reservoir structure and parallel reservoir structure, to three catchments with different underlying surface characteristics. Result indicates that there is a certain relationship between catchment characteristics and model structure. For example, the parallel reservoir structure is suitable for the region with strong permeability and multiple aquifers, the series reservoir structure is suitable for the weak infiltration of bedrock, as the linear structure is suitable for the catchment with stable and smooth runoff variation (Fenicia *et al.* 2014).

Flash floods caused by rainfall of high intensity and short duration occur frequently in the semi-humid or semi-arid regions that within a size of 3,000 km^2. For the northwest of China, due to the problems of data shortage and the complex of underlying conditions, the accuracy of flood forecasting is often too low to fulfill the needs of flood warning (Kan *et al.* 2017). A single model structure is unlikely to precisely describe a diverse range of runoff generation mechanisms (Castillo *et al.* 2003; Clark 2008). So far, the application of conceptual models for flood simulation in semi-humid and semi-arid regions still requires further study (Pilgrim *et al.* 1987). Therefore, it is worth investigating the application of existing models. It is also important to explore which kind of rainfall-runoff model could accommodate the runoff generation mechanisms for this region, as well as the main impact factors for model performance. In this study, six conceptual hydrological models were tested in several semi-humid or semi-arid catchments that located in Shannxi Province, China. The difference in runoff generation for spatial proximity catchments with different climate and geographic properties was investigated and compared. According to the model performance, the applicable of models in the study region will be discussed based on the simulation result, catchment characteristics, model structure and the observed data.

The objective of this work is to investigate the influence of precipitation and geographic conditions on model performance. The study catchments are described in the section Study area, followed by the models and methodology in the section Models and methodology. Results and discussion present the simulated model performances and the discussion of the results. The conclusion is summarized in Conclusion.

STUDY AREA

Three typical semi-humid and semi-arid catchments: Maduwang, Banqiao and Zhidan, which are all sub basins of the Yellow River Basin, were selected to investigate the model performance using different types of models (Figure 1). These three catchments are located in Shannxi province of China with varying catchment characteristics and climate conditions. The meteorological characteristics for the study catchments are listed in Table 1. The distribution of

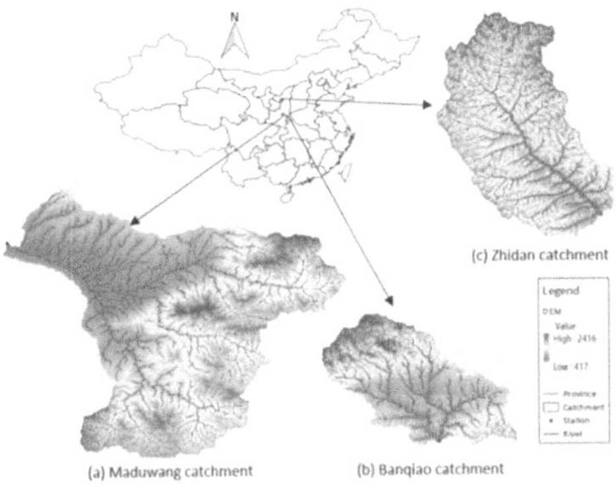

Figure 1 | Location of the selected catchments in Shannxi Province, China.

vegetation types (Table 2) is also calculated based on the global 1 km land cover data provided by the University of Maryland (Friedl *et al.* 2002).

Maduwang catchment has the largest size of 1,604 km^2. The annual precipitation for this catchment is about 630 mm and shows gradually increasing trend from north to south. The mountains and valleys in this area are quite steep with a good coverage of vegetation. The valleys in the hilly areas are extremely developed with large cutting depth, broken terrain and poor vegetation. The plain is flat and the soil is fertile, which is suitable for crop cultivation. The average annual precipitation in the Banqiao River Catchment is around 730 mm with an area of 502 km^2. The terrain is high in the northwest, low in the southeast, most hilly areas contain clayey and sand layer clay, rock type include schist and phyllite, and cultivated land is distributed in river valleys. Zhidan catchment has only 510 mm precipitation per year, and covers a drainage area of 777 km^2. The geographic features of Zhidan can be roughly divided into three types of landforms: valley terraces, beam-shaped gullies and earth-rock mountains. The topographic terrain is distributed with mountains, canyons and barren beaches. The slope varies greatly in Zhidan catchment and the soil erosion is serious because of the poor vegetation cover.

The determination of the humid condition is usually based on water balance, while Liu suggested that the aridity index could be used for natural zonation division in Shaanxi province when the requiring data for calculating water balance is not available (La Vigna *et al.* 2016). The aridity index, which represents the ratio of long-term potential evapo-transpiration to precipitation, is about 1.4 for Maduwang, 1.6 for Banqiao and 2.0 for Zhidan, respectively. According to Liu's study on natural zonation in the Shannxi province, based on the aridity index and annual precipitation, Maduwang and Banqiao catchments are identified as semi-humid region, while Zhidan is regarded as semi-arid region due to the low precipitation and high dryness fraction.

MODELS AND METHODOLOGY

Six conceptual hydrological models are tested in the study catchments: the Sacramento model (named as M1)

Table 1 | Meteorological characteristics for the three catchments

Catchment	Area (km^2)	Annual precipitation (mm)	Annual potential evapotranspiration (mm)	Average temperature (°C)	Average elevation (m)	Annual runoff (mm)	Aridity index
Maduwang	1,601	630.9	1,186	13.3	1,027	307	1.4
Banqiao	493	729.4	1,300	12.8	880	174	1.6
Zhidan	774	509.8	1,556	7.8	1,417	41.6	2

Table 2 | Distribution of vegetation types for the three catchments

Catchment	Evergreen coniferous forest	Broadleaved deciduous forest	Mixed forest	Forest	Forest steppe	Shrubbery	Grass land	Crops
Maduwang	0.163	0.095	0.026	0.219	0.119	0.002	0.175	0.200
Banqiao	0.017	0.001	0.006	0.295	0.390		0.137	0.154
Zhidan					0.004	0.473	0.523	

(Anderson *et al.* 2006), TOPMODEL model (M2) (Beven 1997), Xin'anjiang model (M3) (Zhao 1992), Infiltration-Excess runoff model (M4), first Infiltration-Excess then Saturation-Excess runoff model (M5), and the Xin'anjiang model combined with Infiltration-Excess runoff (M6). The Sacramento model, TOPMODEL and Xin'anjiang model are very popular rainfall-runoff models that applied widely across the world. In Infiltration-Excess runoff model, the Green-Ampt equation is used to describe the infiltration process (Winchell *et al.* 1998). It assumed that the infiltration-excess process occurs on the subsurface and the subsurface runoff is not considered in the model. The infiltration capacity distribution curve is introduced to account for the inhomogeneity of infiltration process and underlying surface. The first Infiltration-Excess then Saturation-Excess runoff model is a two runoff-components model and the runoff concentration calculation is similar to Xin'anjiang model that uses the concept of linear reservoirs. The last model is the combination of Xin'anjiang and Infiltration-Excess runoff model, compared with the traditional Xin'anjiang model, this model contains the infiltration-excess process for the unsaturated zones. It should be noted that only the infiltration-excess surface runoff is generated in the unsaturated area, the interflow and subsurface flow are not considered in the model. More details about the description of the models can be found in Huang (Huang *et al.* 2016).

Firstly, the sensitivity of model parameters was analyzed for the six conceptual models. The most sensitive parameters for each model were selected according to the previous studies to reduce the total number of parameters required in parameterization (Perrin *et al.* 2001). The simplex method is a classical optimization algorithm (Nelder & Mead 1965). Thanks to characteristics of being easy-to-code, fast in searching speed and an excellent performance in approaching optimal solution, the method has been widely used in the research of hydrology and water resources. Compared with SCE-UA algorithm (Guangyuan *et al.* 2016), the simplex method is relatively simple and requires fewer calculations for finding the optimal parameters. In addition, the simplex method is not limited by the number of variables; the increasing of the parameter dimension does not lead to a significant increase in computation. However, the simplex method is still in the stage of empirical method. It is very useful to solve the problem that contains only a few variables. When the number of variables is increasing to a specific value, the accuracy of model calibration using simplex method is generally lower than using the SCE-UA algorithm. Here only the sensitive parameters were considered to be calibrated based on historical data, thus the simplex method was selected to identify the model parameters.

The following four performance criteria were evaluated for study catchments.

The relative error of runoff depth (ΔR):

$$\Delta R(\%) = \left|\frac{R_{sim} - R_{obs}}{R_{obs}}\right| \times 100\% \tag{1}$$

The relative error of peak flow (ΔQ_p):

$$\Delta Q_p(\%) = \left|\frac{Q_{psim} - Q_{pobs}}{Q_{pobs}}\right| \times 100\% \tag{2}$$

where R_{sim} and R_{obs} refer to the simulated and observed runoff depth, and Q_{psim} and Q_{pobs} refer to the simulated and observed peak flow.

The Nash-Sutcliffe efficiency (NS) (Nash & Sutcliffe 1970)

$$NS = 1 - \frac{\sum_{i=1}^{n} [Q_{sim}(i) - Q_{obs}(i)]^2}{\sum_{i=1}^{n} [Q_{obs}(i) - \bar{Q}_{obs}]^2} \tag{3}$$

where $Q_{sim}(i)$ and $Q_{obs}(i)$ are the simulated and observed discharge at given time i, and $\bar{Q}_{obs}$ is the average discharge over the whole period.

The difference of simulated peak flow appearance time T_{qpsim} and the real flow appearance time T_{qpsim} is also calculated:

$$\Delta T_{qp} = |T_{qpsim} - T_{qpobs}| \tag{4}$$

According to the standards of 'Forecasting norm for hydrology intelligence' in China (Fan *et al.* 2016), the accuracy of flood prediction is evaluated based on the qualification rate of runoff depth, peak flow and flood peak appearance time. Therefore, the Equations (1), (2) and (4) are transformed into a binary function in order to make the forecasting result more distinct:

$$QU_R = \begin{cases} 1, & \Delta R(\%) \leq 20\% \\ 0, & \Delta R(\%) > 20\% \end{cases} \quad (5)$$

$$QU_p = \begin{cases} 1, & \Delta Q_p(\%) \leq 20\% \\ 0, & \Delta Q_p(\%) > 20\% \end{cases} \quad (6)$$

$$QU_{T_{qp}} = \begin{cases} 1, & \Delta T_{qp} \leq 3h \\ 0, & \Delta T_{qp} > 3h \end{cases} \quad (7)$$

where QU_R, QU_p and $QU_{T_{qp}}$ denote the qualification rate of runoff depth, peak flow and peak appearance time. Value of 1 indicates qualified and 0 means unqualified, respectively.

The magnitudes of floods differ from catchments, for dry catchment the total runoff depth might be small. The error of runoff depth and the Nash-Sutcliffe efficiency might not be appropriate to evaluate the ability of models for flood forecasting. The most focus of flood forecasting in the semi-humid and semi-arid region of China is the predicting of peak flow value and flood leading time. In this study, the combination of qualification rate for runoff depth and peak flow was taken as objective function for model calibration using the simplex method.

RESULTS AND DISCUSSION

The hourly records of precipitation and discharge for the three catchments is only observed during the flood season. Therefore, the models were simulated based on flood events. A total of 12 flood events in Maduwang, 13 floods in Banqiao and 15 floods in Zhidan were selected. Figure 2 shows one of the simulated hydrographs (Mdduwang: 070713; Banqiao: 030824; Zhidan: 100618) for the study catchments using different types of models. It has been detected from the flow curves that Maduwang catchment is mainly occupied by saturation-excess runoff. But there are also several narrow floods along with sharply increasing and decreasing flow, which indicates that infiltration-excess surface runoff also has relatively important impact on flood. The infiltration-excess process occurs more often in Banqiao than in Maduwang, as Banqiao has very steep hydrographs for most of the flood events and the flow curve is sharp with a relatively high peak value. The hydrograph of Zhidan fluctuate obviously indicates that it is a typical catchment with infiltration-excess runoff mechanism.

Result of Maduwang

For the 12 flood events from Maduwang, seven of them were used to calibrate the models and the remains were used for

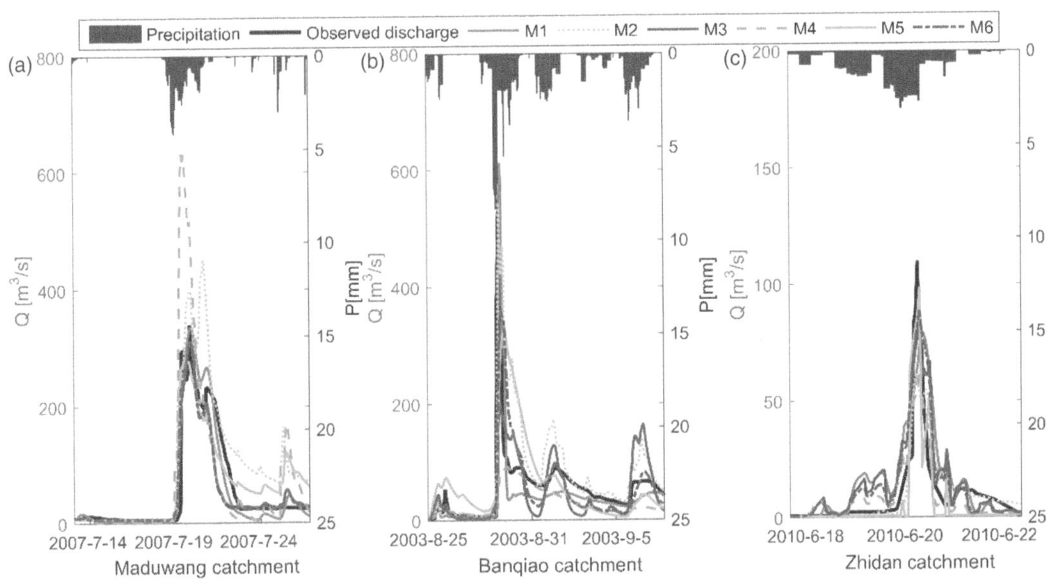

Figure 2 | Simulated hydrograph for three study catchments using different models.

validation. Table 3 presents the model simulation results for both calibration and validation period. Meanwhile, the result of Nash-Sutcliffe efficiency and qualification rate of peak flow were shown in Figure 3.

Result shows that the Xin'anjiang model combined with Infiltration-Excess runoff has the best performance, followed by Xin'anjiang model. The first Infiltration-Excess then Saturation-Excess runoff model shows poorer behavior in the qualification rate of runoff depth compared with the Sacramento model and Xin'anjiang model, but it obtains the best performance in the qualification rate of peak flow. Sacramento model and Infiltration-Excess runoff model are comparable in terms of obtaining reasonable qualification rate of peak flow, while Sacramento model performs better in reduction of the error of runoff depth. TOPMODEL was not able to perform well for all performance criteria in this catchment. It can be seen clearly from Figure 3 that the performance of validation period is generally higher than the calibration period both for NS efficiency and the simulation of peak value. It is because the flood events of the calibration period are more complex than the validation period, some of the steep floods could not be reproduced reasonable by the models that leads to a decreasing of model performance. For the validation period, the dynamic behaviors of the floods are similar to some flood events that used for calibration.

The reason for poor performance of TOPMODEL was investigated. TOPMODEL model is a semi-distributed model that based topography. The interflow is calculated as the negative exponential function of the mean depth of groundwater, and the surface flow is only generated until the groundwater level rises to the surface. In semi-humid region, such as Maduwang catchment, the groundwater level is too deep to reach the surface. The generated surface

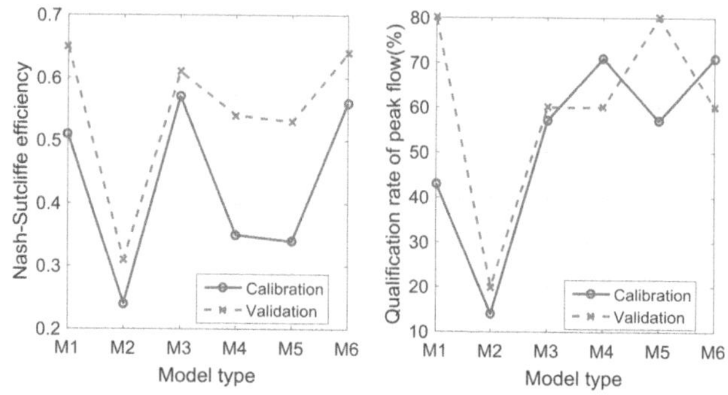

Figure 3 | Model performance for Maduwang catchment.

flow and interflow is small that leads to the underestimation of peak flow. The model performance of Xin'anjiang combined with Infiltration-Excess runoff model is prior to the traditional Xin'anjiang indicating that infiltration-excess runoff element plays an important role in the runoff simulation for Maduwang catchment.

Result of Banqiao

Table 4 shows the model performance of Banqiao catchments for both calibration (eight flood events) and validation (five flood events) period. The results show that the qualification rate of runoff depth is very low for all the models. As mentioned earlier, that peak flow is the main focus of flood prediction and only the qualification rate of

Table 4 | Comparison of model performance for Banqiao catchment

Model type		M1	M2	M3	M4	M5	M6
Qualification rate of peak flow (%)	Calibration	38	38	50	38	50	50
	Validation	20	20	20	20	40	40
Qualification rate flood peak appearance time (%)	Calibration	63	75	50	50	50	50
	Validation	20	40	60	60	80	80

Table 3 | Comparison of model performance for Maduwang catchment

Model type		M1	M2	M3	M4	M5	M6
Qualification rate of runoff (%)	Calibration	71	14	71	29	29	71
	Validation	60	60	80	60	60	80
Qualification rate flood peak appearance time (%)	Calibration	43	14	57	71	57	71
	Validation	80	20	60	60	80	60
Nash-Sutcliffe efficiency	Calibration	0.65	0.31	0.61	0.54	0.53	0.64
	Validation	0.51	0.24	0.57	0.35	0.34	0.56

peak flow and occurring time have been taken as objective function for model calibration. It can be found that for all six models, only 4–6 flood events could achieve qualified peak flow simulation. In general, the model performance for Banqiao catchment is rather poor. The drainage area of Banqiao catchment is small and the mountain torrents runs very fast to the outlet with a short leading time. Thus, the qualification rate of peak flow appearance time is also a critical indicator. Figure 4 shows the simulated discharge series of flood event 030824 for Banqiao using TOPMODEL. It can be seen from the curve that in spite of the big error in discharge volume, TOPMODEL is able to obtain perfect prediction of peak flow in terms of magnitude as well as the appearance time.

Result of Zhidan

For a total of 15 selected flood events, 11 floods were used for model calibration and the remains used for validation. The simulation result (Table 5) demonstrates that Infiltration-Excess runoff model and the Xin'anjiang model combined with Infiltration-Excess runoff model have the best performance in the simulation of peak value, while the latter shows more skill in the simulation of runoff depth. It indicates that even in an extremely dry region, only considering the surface runoff mechanism in runoff simulation may not be appropriate. For Infiltration-Excess runoff model, the reason for less effective performance than the Xin'anjiang model combined with Infiltration-Excess runoff model is also investigated. The runoff is separated into surface runoff and underground water in Infiltration-Excess runoff model. The surface runoff is generated once the rainfall intensity is greater than the infiltration capacity of soil, and the groundwater is generated when the infiltration amount exceeds the deficit of soil water. Due to the mountain terrain, the surface runoff generation and concentration process could be performed in a short time. However, the model assumes that for some places, the infiltration exceeds the deficit of soil water. This causes the instability of model performance, while some simulated floods match well with the observations, some of them lead to serious overestimating or underestimating of peak flow.

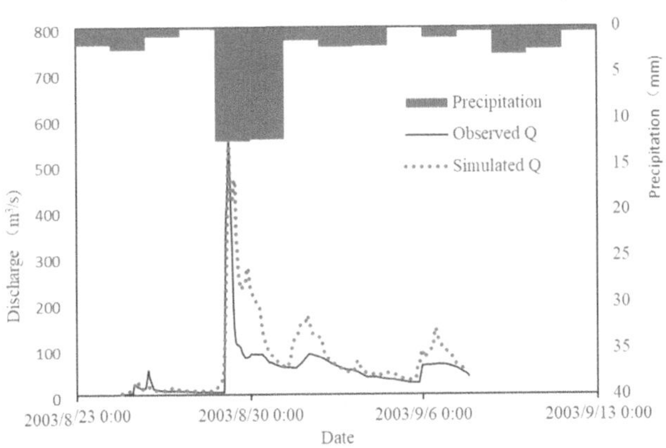

Figure 4 | Hydrograph for the flood event 030824 of Banqiao catchment using TOPMODEL.

Table 5 | Comparison of model performance for Zhidan catchment

Model type		M1	M2	M3	M4	M5	M6
Qualification rate of peak flow (%)	Calibration	27	9	27	27	27	36
	Validation	25	25	25	50	25	25
Qualification rate flood peak appearance time (%)	Calibration	64	64	91	82	64	80
	Validation	25	25	25	50	50	50

Results for different catchments

From the comparison of simulation results for different catchments we can conclude that for the selected models, Maduwang shows the best model performance, followed by Banqiao, and the model simulation for Zhidan could not obtain reasonable performance for most of the flood events. In general, the model performance for these three catchments is not satisfied and could not fulfil the standard requirements of flood forecasting. The poor model performance for the study area might be due to the particular runoff characteristics. Under the long-term drought condition, the runoff generation is dominated by infiltration-excess runoff at the beginning of the rainfalls. If the soil is porous and well developed, the saturation-excess runoff is likely to occur subsequently. However, if the soil is not well developed and packed together, perhaps only infiltration-excess runoff occurs during rainfall. Moreover, the sub-daily rainfall data is normally measured every 6 h for the study area, and the hourly inputs for modeling is based on the mean value of the observed sub-daily rainfall. This results in the reduction of rainfall peak intensity. High temporal resolution of rainfall is required for infiltration-excess runoff model to accurately reproduce the discharge series. The low temporal resolution of rainfall record to some extent causes the low accuracy of model simulation results.

Impacts of topographic features for model performance

Both landscape factors and climate conditions contribute to runoff mechanisms for a specific catchment (Buda *et al.* 2010). In this work, the influence of climate conditions and underlying characteristics to the model performance are investigated. The three catchments are geographically close to each other as they are located in the same province. The climate conditions are relatively different as mentioned in the catchment description but with the same feature of clearly dry and wet seasons. The rainfall-runoff correlation for the catchments is plotted in Figure 5. Here, p denotes areal precipitation, pa denotes antecedent precipitation and R is the observed runoff depth. According to the assumption of saturation-runoff mechanism, all the precipitation produces runoff after the soil is saturated, so the correlation curve close to the diagonal direction indicates the runoff generation is dominated by saturation-runoff mechanism. It can be detected from the figure that for Banqiao catchment, while the rainfall amount of this area is similar to that of Maduwang, the proportion of saturation-runoff is obviously lower than Maduwang. It indicates that variation in precipitation is not the only reason for the great differences in model performance.

It is well known that the spatial and temporal dynamics of runoff generation area highly depends on the landscapes. Topographic index, which is highly related to the runoff concentration area and slope, is usually used for measuring the topographic conditions:

$$T_s = ln\frac{\alpha}{\tan\beta} \tag{8}$$

where T_s is the topographic index, α is runoff concentration area and β is the slope. Figure 6 shows the distribution of topographic index and the corresponding catchment area.

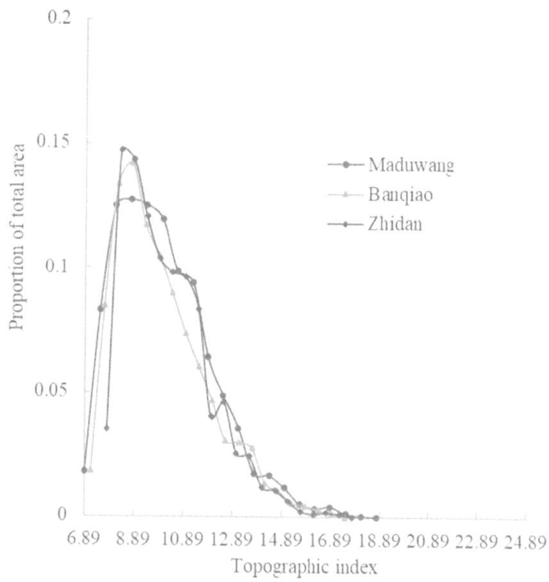

Figure 6 | Relationship between topographic index and the corresponding ratio of catchment area.

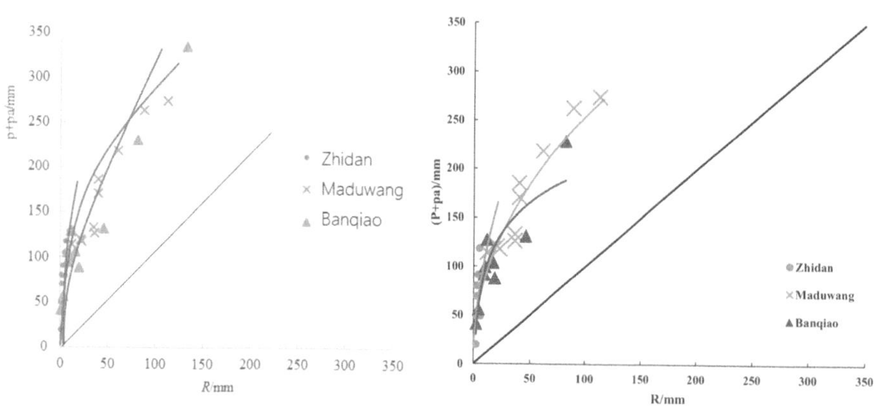

Figure 5 | Rainfall-runoff correlation for the study catchments.

The distribution curve for Maduwang catchment deflects to the right side and over half of the area has the topographic index between 8.8 and 10.8, indicating this catchment has higher proportion of high topographic index area than the other two catchments. Maduwang has the greatest topographic index, followed by Zhidan, while Banqiao has the smallest topographic index. The distribution of soil content that reflected by topographic index could represent the wetness condition of catchment. However, topographic index seems not significantly different between the catchments. Therefore, the types and distribution of vegetation is analyzed to further investigate the impact of surface underlying to runoff generation. As shown in Table 2, Maduwang has various types of well-developed vegetation. The proportion of forest for Banqiao is much less than Maduwang. In Zhidan, most of the area is covered by shrubs and grass. The reason for poorer model performance of Banqiao than Maduwang might be due to the joint influence of topographic index and land cover. The saturation-excess runoff is more likely to occur in Maduwang which has relatively high soil moisture and well-developed vegetation. This is also the reason for the slowness of the flood process and relatively long duration in this catchment.

The average slope and channel gradient are estimated based on digital elevation model (DEM) data. Table 6 shows the average slope and Figure 7 shows the distribution curve for slope. The mean slope of Maduwang is much larger than the other two catchments, while the slope of Banqiao is about 1^0 higher than Zhidan. When the runoff accumulation areas are same, the larger the slope is, the smaller the topographic index is. The sorting of slope for the three catchments is consistent with the result for topographic index, indicating that the effective runoff generation area plays an important role in reflecting the topographic index of the river catchment. Compared to Banqiao, Maduwang has a greater value in the mean of slope degree for the whole catchment, but has a smaller value if it only focuses on the slope degree for the river channel. It is found that Maduwang has rather complex terrain; the high slope degree is distributed in the upstream region that belongs to the hilly and mountainous area. Banqiao has a small catchment size with huge slope in the river channel, which implies the catchment is lack of capacity for water storage. The infiltration-excess runoff is more likely to occur in this catchment at the beginning of rainfall.

Table 6 | Comparison of slope degree for three catchments

Catchment	Mean slope for whole catchment (°)	Mean slope for river channel (°)
Maduwang	17.01	8.17
Banqiao	14.85	8.87
Zhidan	15.72	4.53

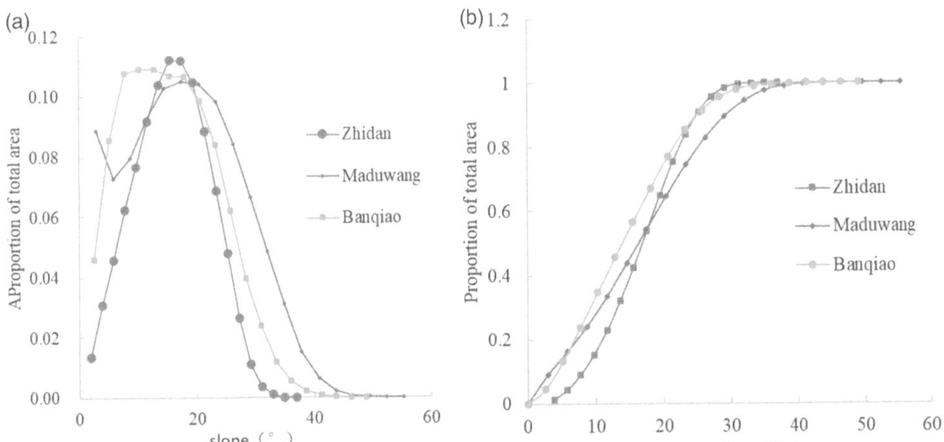

Figure 7 | Distribution of slope (a) and the cumulative distribution curve of slope (b) for the three catchments.

CONCLUSION

In this study, the influence of precipitation and topographic conditions on the model simulation results were tested on three catchments using six different models. The results show that for the region where the runoff generation mechanism was dominated by the infiltration-excess runoff, the model that incorporated infiltration-excess runoff mechanism could obviously lead to better model performance than if only the saturation-excess runoff had been considered. In the semi-humid region, most of the traditional model could not perform reasonably because of the low capability in simulating the infiltration-excess runoff process. The model with flexible structure, such as the first Infiltration-Excess then Saturation-Excess runoff model and the combination of Xin'anjiang with Infiltration-Excess runoff model, shows some advantages in the model simulation for the place that both runoff generation mechanisms exist.

Results indicate the topographic characteristics and climate conditions of a catchment strongly affect the rainfall-runoff process. Floods are relatively similar and easier to be simulated in the region with well-developed vegetation and simple terrains. The influence of slope for a river channel is higher than that mean slope for the whole catchment. It has been detected from this work that for a particular catchment, the higher the topographic index is, the greater the possibility of saturation-excess runoff generation is.

It can be found for the study catchments that model simulation results are impacted by several factors. Not only the structure of the model, but also the accuracy of the observations and the underlying conditions could affect the model performance. Therefore, for flood simulation in semi-humid or semi-arid regions, sufficient analysis has to be taken before determining whether the model is applicable. Moreover, when the existing models could not adapt to the complex runoff generation process, adjusting the model structure by adding some corresponding modules to accommodate all possibilities could be a good choice for flood simulation in mixed runoff generation regions.

DATA AVAILABILITY

The precipitation and discharge data used to support the findings of this study were supplied by Shaanxi Hydrology Bureau under license and so cannot be made freely available. Requests for access to these data should be made to Long Sun (sunlongmwr@163.com).

FUNDING STATEMENT

This research was funded by National Key Research and Development Program of China (2018YFC1508101), National Natural Science Foundation of China (51679061), the Non-profit Industry Financial Program of MWR of China (201401008) and Non-profit Industry Financial Program of Ministry of Water Resources, China (201001045).

DATA AVAILABILITY STATEMENT

All relevant data are included in the paper or its Supplementary Information.

REFERENCES

Adnan, R. M., Liang, Z. & Kisi, O. 2020 Discussion of 'performance enhancement of a conceptual hydrological model by integrating artificial intelligence' by Ahmet Ali Kumanlioglu and Okan Fistikoglu. *Journal of Hydrologic Engineering* **25** (9).

Alhassan, A. & Jin, M. G. 2020 Evapotranspiration in the tono reservoir catchment in upper east region of Ghana estimated by a novel TSEB approach from ASTER imagery. *Remote Sensing* **12** (3), 569.

Anderson, R. M., Koren, V. I. & Reed, S. M. 2006 Using SSURGO data to improve Sacramento Model a priori parameter estimates. *Journal of Hydrology* **320** (1–2), 103–116.

Beven, K. 1997 TOPMODEL: A critique. *Hydrological Processes* **11** (9), 1069–1085.

Buda, A. R., Kleinman, P. J. A., Srinivasan, M. S., Bryant, R. B. & Feyereisen, G. W. 2010 Factors influencing surface runoff generation from two agricultural hillslopes in central Pennsylvania. *Hydrological Processes* **23** (9), 1295–1312.

Castillo, V. M., Gomez-Plaza, A. & Martinez-Mena, M. 2003 The role of antecedent soil water content in the runoff response of semiarid catchments: a simulation approach. *Journal of Hydrology* **284** (1–4), 114–130.

Chao, L., Zhang, K., Li, Z., Wang, J., Yao, C. & Li, Q. 2019 Applicability assessment of the CASCade two Dimensional SEDiment (CASC2D-SED) distributed hydrological model for flood forecasting across four typical medium and small

watersheds in China. *Journal of Flood Risk Management* **12** (14), 23–31.
Chirico, G. B., Grayson, R. B. & Western, A. W. 2003 A downward approach to identifying the structure and parameters of a process-based model for a small experimental catchment. *Hydrological Processes* **17** (11), 2239–2258.
Clark, M. P. 2008 Framework for understanding structural errors (FUSE): a modular framework to diagnose differences between hydrological models. *Water Resources Research* **44** (12), 1–13.
Clark, M. P., Rupp, D. E., Woods, R. A., Tromp-van Meerveld, H. J., Peters, N. E. & Freer, J. E. 2009 Consistency between hydrological models and field observations: linking processes at the hillslope scale to hydrological responses at the watershed scale. *Hydrological Processes* **23** (2), 311–319.
Dunne, T. & Black, R. D. 1970 Partial area contribution to storm runoff in a small new england watershed. *Water Resources Research* **6** (5), 1296–1311.
Euser, T., Hrachowitz, M., Winsemius, H. C. & Savenije, H. H. G. 2015 The effect of forcing and landscape distribution on performance and consistency of model structures. *Hydrological Processes* **29** (17), 3727–3743.
Fan, F., Paiva, R., Collischonn, W. & Ramos, M. H. 2016 Hydrological forecasting practices in Brazil. *Flood Forecasting* **18**, 41–66.
Fenicia, F., Kavetski, D., Savenije, H. H. G., Clark, M. P., Schoups, G., Pfister, L. & Freer, J. 2014 Catchment properties, function, and conceptual model representation: is there a correspondence? *Hydrological Processes* **28** (4), 2451–2467.
Friedl, M. A., McIver, D. K., Hodges, J. C. F., Zhang, X. Y., Muchoney, D., Strahler, A. H., Woodcock, C. E., Gopal, S., Schneider, A., Cooper, A., Baccini, A., Gao, F. & Schaaf, C. 2002 Global land cover mapping from MODIS: algorithms and early results. *Remote Sensing of Environment* **83** (1–2), 287–302.
Guangyuan, K., Ke, L., Jiren, L., Liuqian, D., Xiaoyan, H., Youbing, H. & Amo-Boateng, M. 2016 Accelerating the SCE-UA global optimization method based on multi-core CPU and many-core GPU. *Advances in Meteorology* **2016** (2016–4–6), 1–10.
Horton, R. E. 1933 The rôle of infiltration in the hydrologic cycle. *Eos Transactions American Geophysical Union* **14** (1), 1–9.
Hrachowitz, M., Savenije, H. H. G., Bloschl, G., McDonnell, J. J. & Cudennec, C. 2013 A decade of predictions in ungauged basins (PUB)a review. *Hydrological Sciences Journal-Journal Des Sciences Hydrologiques* **58** (6), 1198–1255.
Huang, P., Li, Z., Yao, C., Qiaoling, L. & Meichun, Y. 2016 Spatial combination modeling framework of saturation-excess and infiltration-excess runoff for semihumid watersheds. *Advances in Meteorology* **2016**, 1–15.
Kan, G. Y., He, X. Y., Ding, L. Q., Li, J., Liang, K. & Hong, Y. 2017 Study on applicability of conceptual hydrological models for flood forecasting in humid, semi-humid semi-arid and arid basins in China. *Water* **9** (10), 25.
Kavetski, D., Kuczera, G. & Franks, S. W. 2006 Calibration of conceptual hydrological models revisited: 1. Overcoming numerical artefacts. *Journal of Hydrology* **320** (1–2), 0–18.
La Vigna, F., Hill, M. C., Rossetto, R. & Mazza, R. 2016 Parameterization, sensitivity analysis, and inversion: an investigation using groundwater modeling of the surface-mined Tivoli-Guidonia basin (Metropolitan City of Rome, Italy). *Hydrogeology Journal* **24** (6), 1–19.
Liu, Z., Wang, Y., Xu, Z. & Duan, Q. 2017 *Conceptual Hydrological Models* **12** (3), 34–45.
Nash, J. E. & Sutcliffe, J. V. 1970 River flow forecasting through conceptual models part I – a discussion of principles. *Journal of Hydrology* **10** (3), 282–290.
Nelder, J. A. & Mead, R. 1965 A simplex method for function minimization. *The Computer Journal* **7** (4), 308–313.
Perrin, C., Michel, C. & Andréassian, V. 2001 Does a large number of parameters enhance model performance? comparative assessment of common catchment model structures on 429 catchments. *Journal of Hydrology* **242** (3), 275–301.
Pilgrim, D. H., Chapman, T. G. & Doran, D. G. 1987 Problems of rainfall-runoff modelling in arid and semiarid regions. *International Association of Scientific Hydrology Bulletin* **33** (4), 379–400.
Singh, V. P. 1997 Computer models of watershed hydrology. *Water Resources Publications*, Highlands Ranch, CA, USA, pp. 443–476.
Winchell, M., Gupta, H. V. & Sorooshian, S. 1998 On the simulation of infiltration- and saturation-excess runoff using radar-based rainfall estimates: effects of algorithm uncertainty and pixel aggregation. *Water Resources Research* **34** (10), 2655–2670.
Winter, T. C. 2010 The concept of hydrologic landscapes. *Journal of the American Water Resources Association* **37** (2), 335–349.
Zhao, R. 1992 The Xinanjiang model applied in China. *Journal of Hydrology* **135** (1–4), 371–381.

First received 6 July 2020; accepted in revised form 10 October 2020. Available online 23 October 2020

Evaluation of water security based on capacity for socio-economic regulation

Lian Tang, Weibing Zhang, Zixi Liu and Yarong Qi

ABSTRACT

From the perspective of social and economic regulation, the development trend of regional water security in different periods and under different regulation strengths is evaluated by using the fuzzy set pair analysis method. The results show: the degree of regional water resources security changes from insecurity in the pilot period of water-saving society construction to basic security, and then to security in the planning period, with the enhancement of economic and social regulation. The order of importance of each index is C19 > C1 = C5 > C18 > C14 > C10 > C13 > C3 = C4 > C8 > C7 > C15 > C16 > C12 > C2 > C11 > C17 > C6 > C9, the long-term shortage of water resources is the most important factor, the contribution rate is 60%, but the index weight of economic and social coordination ability is large, and its rapid change causes the improvement of water resources security; the contradiction between the regulation intensity of social and economic indicators in the planning year and the demand for water resources is not matched, and the degree of water security is mainly restricted by the shortage of regional water resources.

Key words | comprehensive evaluation, fuzzy set pair analysis, social and economic regulation, water resources security

Lian Tang† (corresponding author)
Weibing Zhang†
Zixi Liu
Yarong Qi
School of Civil and Water Conservancy Engineering of Ningxia University,
Yinchuan 750021,
Ningxia,
China
and
Engineering Research Center of Ministry of Education for Efficient Utilization of Modern Agricultural Water Resources in Dry Areas,
Yinchuan 750021,
China
E-mail: *nxdxtl@126.com*

Lian Tang†
Weibing Zhang†
Innovation Team of Ministry of Education for Efficient Utilization of Modern Agricultural Water Resources in Dry Areas,
Yinchuan 750021,
China

†Both authors contributed equally to this work.

HIGHLIGHTS

- The important role of social and economic regulation and control measures in water shortage was discussed.
- The influencing factors on water resource shortage were discussed.
- The evaluation of water resource safety can be judged by comprehensive connection degree.
- The contribution of factors in different years to regional water resources security was analyzed.
- Strengthening economic and social control measures is conducive to solving the problems of regional water resource shortage.

INTRODUCTION

Population growth, socio-economic development and urbanization have aggravated climatic effects in arid areas, resulting in a series of water resources problems, such as water pollution, fragile aquatic ecosystems and frequent drought and flood disasters. At the same time, the implementation of water-saving interventions, stringent water resources management, sustainable development and other adaptive measures (Zhengwei *et al.* 2016; Xia & Bing 2018) have to a certain extent improved and enhanced the robustness of water resources systems in China. Adapting to changing social and environmental factors to improve the security of water resources has become a research focus.

doi: 10.2166/ws.2020.287

Different studies have considered different definitions of water security. Some scholars have applied a relatively narrow definition, considering security of supply, balancing supply and demand, water resources carrying capacity and environmental carrying capacity (Page 2001; Yiping *et al.* 2013; Wen *et al.* 2014). The assessment of water security has in recent years become more important with increasing social changes. With clear objectives for the quantity and quality of water resources, and under pressure of limited water resources, regulation is vital for consolidating environmental objectives with economic and social adaptation measures (Handmer *et al.* 1986).

Water security is a complex social issue, and includes aspects of economic production and human lifestyle (Zhifei 2017). In addition to the quantity and quality of water resources, water security should also consider (Jun & Wei 2016) the utilization and management of water resources. Water security is impacted by social changes (Jun *et al.* 2011) such as developments in science and technology and level of social development. It is emphasized (Yirong & Jiancang 2014; Jun *et al.* 2016) that coordination between the water resources system as a complex socio-economic and environmental system and an evaluation index system should reflect the binary nature of the natural and socio-economic system. Water security based on the most stringent water management system (Junyuan *et al.* 2016) is the embodiment of regulating water security through social management.

Water security should in addition to the water resources themselves include social and economic changes, scientific and technological development and management and control of water resources. The negative impacts of climate change in arid areas may lead to increasing frequency of floods and droughts, water shortages and increasing risk of deterioration of the aquatic environment, thereby resulting in a serious water security crisis (Jun *et al.* 2011). These factors could result in water constraints persisting for a long time. When water shortages and resulting environmental problems cannot entirely be overcome by fully using the inherent potential of water resources, adaptation to more efficient use of water resources is required (Ohlsson 1998; Qiting 2017). Socio-economic regulation and control of capacity can be used to enhance water security (Ohlsson 2000) through the mobilization of secondary resources (social resources), enhancing socio-economic adaptability (Huaiwen *et al.* 2011), alleviating pressure on water resources and restoring system function processes (Huaiwen *et al.* 2011; Liping & Deshan 2014).

The limited water resources of arid areas have a great impact on water security and this is difficult to change in the short term. Through the enhancement of active coordination between society, economy and environment, limitations of the water resource can be compensated for and the water needs of social development can be met.

The evaluation of water security should include social and economic regulation. Water security should encompass the degree to which the water quantity and quality needs of society can be met and the probability of guaranteeing sustainable development of the economy, society and ecology. 'Control efficiency' is based on the degree and effect of social and economic regulation and adjustment, such as (Yirong & Jiancang 2014) implementation of water-saving interventions, stringent water resources management policies, social investment in environmental management and protection, comprehensive consideration of changes in social factors such as urbanization and population growth and consideration of comprehensive water resources control measures. This paper selects the water-saving society pilot construction period of 2006–2013 for research historical data, and compares the planning data from 2018 to 2020 to explore the impact of different regulation intensities on water security and its change trend.

MATERIALS AND METHODS

Definition of the water security evaluation index

Ningxia in northwest China encompasses arid and semi-arid areas, and often suffers serious water shortages. Water security is the key factor required for sustainable development. In the case of long-term and serious shortage of water resources, the proportion of agricultural water consumption is very large. The agricultural water is rich but its utilization rate is very low. The water utilization coefficient of agricultural irrigation is far lower than the average level of China (Table 1).

To improve the efficiency of agricultural water consumption is to improve the safety and security of the region. The fairness and interests of regional water use are

Table 1 | The basic conditions of the water resource in the studied area

Annual average precipitation	Annual potential evaporation	Per capita annual water availability	The proportion of agricultural water consumption (2013)	The water utilization coefficient of agricultural irrigation (2013)	Urban sewage recycling rate C4 (2013)
305 mm	1,800 mm	634 m^3	87%	0.48	14.6%

greatly affected by agricultural water-saving. Therefore, when selecting indicators, this paper focuses on the agricultural water-saving situation, and does not consider the ecological, industrial and living aspects in detail. By referring to domestic and foreign literature (Min *et al.* 2006; Huaiwen *et al.* 2011; Li & Songhong 2014; Liping & Deshan 2014), 19 indices were selected to establish a water safety evaluation index for Ningxia (Table 2).

(1) The basic characteristics of water security are reflected by water resources condition indices C1–C5.
(2) The water security assessment not only includes basic functions of water resources, but also economic, social and environmental service functions, particularly the increases to water security provided by economic, social and environmental regulation. Therefore, indices that reflect coordination between society, economy and the aquatic environment reflect 'the efficiency of active economic and social regulation'. The selection of indicators focused on the implementation of social progress and adjustment capacity C6–C10 and capacity for economic coordination to enhance implementation C11–C14. Finally, indices considering the impacts of coordination with the aquatic environment were selected, namely C15–C19. Because the indicators of coordination ability should not be quantified and compared, the present study selected efficiency and benefit indicators to express the effect of coordination, focusing on the support of urban water-saving technologies and the growth of the city economy.

Determination of security criteria for indicators

According to the meaning of the indicators, relevant literature and the water availability within the study region, water security criteria were categorized into five levels.

(1) Very safe (Level I): The socio-economic factors, aquatic environment and social coordination capabilities are compatible with regional water resources conditions and are very secure.
(2) Moderately safe (Level II): The socio-economic factors, aquatic environment, social coordination capacity and regional water resources conditions are moderately adaptive and safe.
(3) Basic safety (Level III): The socio-economic factors, aquatic environment and social coordination capabilities are in general adapted to regional water resources conditions; however, the system is vulnerable to exceedance of a critical point of safety and security which may result in water insecurity.
(4) Insecure (Level IV): The socio-economic factors, aquatic environment, social control capacity and regional water resources are not coordinated, and water resources cannot support regional economic and social development.
(5) Extremely unsafe (Level V): The water resources and aquatic environment are seriously polluted, and the social economy, aquatic environment and capacity for social control are inadequate. The fundamental problems resulting from water shortages cannot be overcome, resulting in lowered economic and social development.

The specific basic data and evaluation level standards are shown in Table 3.

Water security evaluation model based on entropy weight fuzzy set

The fundamentals of water security assessment include socio-economic factors, water resources and aquatic environment composite system aspects, which are associated with large degrees of uncertainty (Min *et al.* 2006; Li & Songhong 2014). Fuzzy set pair analysis (Jun *et al.* 2016) is a multi-objective decision-making method that introduces the concept of the degree of fuzzy connection to deal with uncertainty based on the theory of set pair analysis.

Table 2 | Water resources safety rating indicator system

Target layer	Criteria layer	Indicator layer	Indicator calculation	Characterization	Unit	Nature
Water security assessment	Urban water resources condition B1	Per capita water resources C1	Total water resources/total population	Measuring the extent of urban water shortages	m^3/person	↑
		Current rainfall C2	The sum of precipitation in the year	Reflecting the precipitation in the urban area	mm	↑
		Water resources development utilization rate C3	Regional water consumption/water resource availability × 100%	Local water resources development and utilization	%	↓
		Urban sewage recycling rate C4	Sewage recycling capacity/ total sewage treatment × 100%	Reflecting the ability of sewage resources	%	↑
		Unit per hm^2 water resources C5	Total water resources/total area	Reflecting the amount of urban water resources	m^3/hm^2	↑
	Social coordination ability B2	Urbanization rate C6	Urban population/total population × 100%	Measuring urbanization levels	%	↑
		Natural population growth rate C7	Annual population increase/total population × 100%	Reflecting the speed of population development	%	↓
		Urban water supply network loss rate C8	(Total water supply – total effective water supply)/ total water supply × 100%	Measuring the water supply efficiency of urban water supply systems	%	↓
		Urban life water-saving equipment penetration rate C9	The number of water-saving appliances used again/the number of reused water appliances × 100%	Reflecting people's understanding of resource crisis	%	↑
		Agricultural water-saving irrigation area proportion C10	Water-saving irrigation area/total area of irrigation area × 100%	Characterizing the optimal allocation of agricultural water	%	↑
	Economic coordination ability B3	Per capita GDP C11	Gross GDP/total population	Measuring regional economic development	Yuan/person	↑
		10,000 yuan GDP water consumption C12	Total water consumption/ total GDP	Economic benefit	m^3	↓
		Agricultural 10,000 yuan increase in water consumption C13	Total water withdrawal/ total agricultural output increased	Reflecting agricultural water efficiency	m^3	↓
		Agricultural water use proportion C14	Agricultural water consumption/total water consumption × 100%	Reflecting the urban industrial water structure	%	↓
	Ecosystem coordination ability B4	Urban green coverage rate C15	Total area of urban green space/total urban area × 100%	Reflecting urban greening levels	%	↑
		Environmental protection investment proportion C16	Environmental protection investment/total investment × 100%	Reflecting the importance of cities to environmental governance	%	↑
		Urban sewage treatment rate C17	Concentrated sewage treatment volume/total sewage volume × 100%	Reflecting industrial water recycling levels	%	↑
		Industrial wastewater discharge compliance rate C18	Industrial wastewater discharge scalar/ industrial wastewater discharge total × 100%	Responsibility to reflect environmental responsibility	%	↑
		Urban drinking water quality compliance rate C19	Drinking water scalar/total reference water × 100%	Measuring water safety	%	↑

Table 3 | The water security assessment set for Ningxia in 2006–2020

Index	2006	2007	2008	2009	2010	2011	2012	2013	2018	2019	2020	Level I Very safe	Level II Moderately safe	Level III Basic safety	Level IV Insecure	Level V Extremely unsafe	Single indicator Weights	Criterion layer weight
C1	175.66	170.19	149.02	134.71	147.09	136.87	166.95	172.21	159.6	160.1	163.5	≥ 3,000	1,700–3,000	1,000–1,700	500–1,000	≤ 500	0.093	0.3136
C2	249	299	250	235	293	284	339	321	389	346	311	≥ 1,000	800–1,000	600–800	400–600	≤ 400	0.028	
C3	95.72	97.15	93.00	90.55	97.53	94.07	98.41	96.12	96.2	96.5	96.6	≤ 30	30–50	50–70	70–90	≥ 90	0.049	
C4	8.01	8.99	10.56	12	13.7	13.98	14.25	14.6	20.2	23	25	≥ 50	30–50	15–30	5–15	≤ 5	0.049	
C5	20,473	20,050	17,770	16,259	17,996	16,896	20,859	21,749	20,923	20,861	20,142	≥ 60,000	30,000–60,000	15,000–30,000	6,000–15,000	≤ 6,000	0.093	
C6	43.10	44.02	44.98	46.10	48.12	49.82	50.67	52.02	58	59.9	61	≥ 80	60–80	40–60	20–40	≤ 20	0.023	0.1864
C7	6.41	9.76	9.69	9.68	9.04	8.97	8.93	8.62	7.78	8.03	8.12	≤ 2	2–4	4–6	6–10	≥ 10	0.046	
C8	19	12	11.9	11.7	8.5	9	8.3	7	7	7	6	≤ 8	8–10	10–15	15–18	18–20	0.032	
C9	38	45	55	70	75	80	83	87	96	97	100	≥ 90	70–90	50–70	30–50	≤ 30	0.015	
C10	1.45	3.20	5.01	5.85	7.20	11.60	16.40	20.17	41.4	47.7	55.2	≥ 50	35–50	20–35	5–20	≤ 5	0.07	
C11	11,710	13,669	17,784	21,346	25,931	32,228	35,950	39,210	50,659	54,500	59,000	≥ 50,000	25,000–50,000	10,000–25,000	4,000–10,000	≤ 4,000	0.027	0.1864
C12	1,098	851	675	541	440	357	298	281	178.6	162	150	≤ 500	500–1,000	1,000–1,500	1,500–2,000	≥ 2,000	0.032	
C13	913	670	576	274	217	189	162	147	115	92	80	≤ 500	500–1,000	1,000–1,500	1,500–2,000	≥ 2,000	0.05	
C14	93.13	92.44	93.19	92.37	91.71	91.02	90.08	87.52	87.2	86.9	86.8	≤ 60	60–70	70–80	80–90	≥ 90	0.078	
C15	16.5	17.4	20.3	25.1	27.6	29.6	30.5	37.9	42.2	43.6	44	≥ 30	20–30	15–20	10–15	≤ 10	0.042	0.3136
C16	4.95	5.14	5.17	5.22	5.23	5.51	6.25	7.82	9.52	9.88	10	≥ 10	7–10	5–7	3–5	≤ 3	0.036	
C17	53.4	54	55.26	71.2	76	84.5	88.2	89	92.2	93.2	95	≥ 90	70–90	50–70	30–50	≤ 30	0.024	
C18	78.2	83.6	91.7	97	97.8	98	98.5	99	97.6	99.1	100	≥ 85	75–85	60–75	45–60	≤ 45	0.081	
C19	94.07	95.48	96.29	99.5	99.9	100	100	100	100	100	100	≥ 99	96–99	93–96	90–93	≤ 90	0.13	

(1) Set pair analysis is based on the system pairing principle and is used to analyze a system consisting of one definite and one uncertain data set, which may be identical, different or contradictory to each other, to deal with and describe the problem of synthesis integration. Fuzzy set pair analysis is based on set pair analysis. Fuzzy logic theory is applied to set pair analysis and is better able to consider the fuzziness of the standard boundary of the grade. Fuzzy logic theory is more objective and allows simpler solutions to uncertain problems. The principles are as follows (Li & Songhong 2014; Yongan & Wensheng 2018).

Using water security as the evaluation target, fuzzy logic theory constitutes a set of water security assessment indicators and safety level standards $H(A\_t,B\_k)$. The evaluation index value consists of $t = 1, 2,\ldots t$, where t is the total number of evaluation indicators, equal to 19. The evaluation index is regarded as a set A_t, with the evaluation level standard set expressed as B_k; $k = 1, 2,\ldots k$, where k is the number of evaluation level standards, equal to 5. According to the basic principle of set pair analysis, the symbol quantization process is performed on H, and when the evaluation index value is compared with the corresponding evaluation level standard, if it happens to be in the kth standard, the quantized symbol 'k', and so on.

(2) According to the set pair principle, the total number of water security evaluation choices is T, where S is the number of indicators that meet the same evaluation level as the evaluation index, called the collection A_t. B_k is common, and is called 'the same'. The number of indicators with different levels of evaluation K-1 is P, called the set A_t, whereas B_k is the opposite, and is called 'the opposite'. The evaluation level differs by level I, level II, and the level of K-2 is recorded as F_1, F_2, $F_{K\text{-}2}$. F is different from the set A_t. B_k is neither shared nor shared, F is for 'the difference'; from this, the K-member relationship of the water security set to H is obtained (Wenjuan et al. 2018):

$$\mu_{A_t-B_k} = \frac{S}{T} + \frac{F}{T}I + \frac{P}{T}J = a + \sum_{1}^{K-2} b_K I_K + cJ \tag{1}$$

In Equation (1), a, b and c represent the identity, difference and opposite components of the sets A_t and B_k, respectively, and meet the conditions $a + b + c \neq 1$; I_1, I_2, $I_3 \ldots I_{k\text{-}2}$ represent the coefficient of the uncertainty component, which has a value of -1 to 1, with a value closer to 1 representing the level to be evaluated; J is the opposite coefficient with a value of $-p$.

The difference uncertainty component coefficient I_k in Equation (1) can be determined by the uniform value method (Wenjuan et al. 2018; Yongan & Wensheng 2018):

$$I_k = \frac{1 - 2k}{(K - 1)} \tag{2}$$

In Equation (1), the closer a is to 1, the more similar are the two sets, whereas the opposite is true for c. The greater the connection value of the two sets, the more likely the indicator value belongs to the level standard.

(3) Calculation of the degree of connection of a single indicator

In the evaluation, B_k is defined as the set B_k of a specific index t-level rating standard; that is, the degree of connection of a single indicator is calculated first. The specific formula reference (Wenjuan et al. 2018) is as follows:

$$\mu_{A_t \sim B_k} = \begin{cases} 1 + 0I_1 + \cdots 0I_{k-2} + 0J & (x_1 \geq s_1) \\ \dfrac{2x_t - s_1 - s_2}{s_1 - s_2} + \dfrac{2s_1 - 2x_t}{s_1 - s_2} I_1 + 0I_2 + \cdots + 0I_{K-2} + 0J & \left(\dfrac{s_1 + s_2}{2} \leq x_t < s_1\right) \\ 0 + \dfrac{2x_t - s_2 - s_3}{s_1 - s_3} I_1 + \dfrac{s_1 + s_2 - 2x_t}{s_1 - s_3} I_2 + \cdots + 0I_{K-2} + 0J & \left(\dfrac{s_2 + s_3}{2} \leq x_t < \dfrac{s_1 + s_2}{2}\right) \\ \cdots & \\ 0 + 0I_1 + \cdots + \dfrac{2x_t - 2s_{K-1}}{s_{K-2} - s_{K-1}} I_{K-2} + \dfrac{s_{K-1} + s_{K-2} - 2x_t}{s_{K-2} - s_{K-1}} J & \left(s_{K-1} \leq x_t < \dfrac{s_{K-2} + s_{K-1}}{2}\right) \\ 0 + 0I_1 + 0I_2 + \cdots + 0I_{K-2} + 1J & (x_t < s_{K-1}) \end{cases} \tag{3}$$

$$\mu_{A_t\sim B_k}=\begin{cases}1+0I_1+\cdots 0I_{k-2}+0J & (x_1\le s_1)\\ \dfrac{s_1+s_2-2x_t}{s_2-s_1}+\dfrac{2x_t-2s_1}{s_2-s_1}I_1+0I_2+\cdots+0I_{K-2}+0J & \left(s_1<x_t\le\dfrac{s_1+s_2}{2}\right)\\ 0+\dfrac{s_2+s_3-2x_t}{s_3-s_1}I_1+\dfrac{2x_t-s_1-s_2}{s_3-s_1}I_2+\cdots+0I_{K-2}+0J & \left(\dfrac{s_1+s_2}{2}<x_t\le\dfrac{s_2+s_3}{2}\right)\\ \cdots & \\ 0+0I_1+\cdots+\dfrac{2s_{K-1}-2x_t}{s_{K-1}-s_{K-2}}I_{K-2}+\dfrac{2x_t-s_{K-1}-s_{K-2}}{s_{K-1}-s_{K-2}}J & \left(\dfrac{s_{K-2}+s_{K-1}}{2}<x_t\le s_{K-1}\right)\\ 0+0I_1+0I_2+\cdots+0I_{K-2}+1J & (x_t<s_{K-1})\end{cases}\tag{4}$$

Equation (3) applies to the attribute where $K>2$ is a positive indicator, where $s_1\ge s_2\ge\ldots\ge s_{K\sim1}$. Equation (4) is applied to the index where the attribute of $K>2$ is negative, and $s_1\le s_2\le\ldots\le s_{K\sim1}$.

(1) The weight of the water safety evaluation index can be determined (Wenjuan *et al.* 2018) by the entropy method. The weight of each indicator is expressed by ω_t. The theory behind the approach is as follows.

1. The initial data of the t indicators of m samples of n different periods constitute a judgment matrix:

$$R=(x_{ij})(i=1,2,\cdots m;j=1,2,\cdots T)\tag{5}$$

x_{ij} is the characteristic value of sample j and index i.

2. The normalized judgment matrix.

The larger the value, the better the indicator:

$$\mu_{ij}=X_{ij}/\max X_{ij}\tag{6}$$

The smaller the value, the better the indicator:

$$\mu_{ij}=\min X_{ij}/X_{ij}\tag{7}$$

3. Determine the entropy of the evaluation index:

$$f_{ij}=\mu_{ij}/\sum_{j=1}^{T}\mu_{ij}\qquad H_i=-\frac{1}{\ln T}\left(\sum_{j=1}^{T}f_{ij}\ln f_{ij}\right)\tag{8}$$

4. Calculate the entropy weight of the evaluation index:

$$W=(\omega_i)_{1\times m}\qquad \omega_i=\frac{1-H_i}{m-\sum_{i=1}^{m}H_i}\tag{9}$$

(2) Calculation of degree of contact

The index corresponding to the set A of the evaluation index system is the evaluation level standard set B of a certain level, and the k-yuan connection degree of $H(A\_t, B\_k)$ of the set pair can be calculated according to the following formula:

$$\begin{aligned}\mu_{A\tilde{B}}&=\sum_{t=1}^{T}\omega_t\mu_{A_t-B_1}\\&=\sum_{t=1}^{T}\omega_t a_t+\sum_{t=1}^{T}\omega_t b_{t,1}I_1+\sum_{t=1}^{T}\omega_t b_{t,2}I_2+\cdots\\&\quad+\sum_{t=1}^{T}\omega_t b_{t,K-2}I_{K-2}+\sum_{t=1}^{T}\omega_t c_t J\\&=f_1+f_2I_1+\ldots+f_{k-1}I_{k-2}+f_kJ,\end{aligned}\tag{10}$$

where f_1 represents the probability that the evaluation target belongs to the I-level standard, f_2 is the probability that the evaluation target belongs to the II-level standard, f_k is the probability that the evaluation target belongs to the K-level standard.

(3) Determine the rating

Calculate the confidence to determine the level of water security, as follows:

$$h_k=(f_1+f_2+f_3+\cdots+f_k)>\lambda,\quad k=1,2,3\ldots K,\tag{11}$$

where h_k is the attribute measure and λ represents the confidence.

In Equation (11), the value of confidence λ should neither be too large nor too small as these would lead to the evaluation results being too conservative or too uncertain, respectively. Therefore, λ should generally fall within [0.5, 0.7], and the current study uses a value of 0.5. The level in which the evaluation sample falls is determined by Equation (11), i.e., the sample belongs to the k level corresponding to h_k.

Data sources

The evaluation system began in 2006 during the comprehensive implementation of water-saving interventions within society and ended in 2013. The basic data of all indicators were obtained from the Ningxia Water Resources Bulletin and Water Resources Survey Bureau 2006–2013 and the Statistical Communiqué of the National Economic and Social Development of Ningxia Hui Autonomous Region 2006–2013. The planning year basic data (2018–2020) were from the Ningxia Water Resources Allocation Planning Report 2016–2020. In 2018, the actual data will be used for statistics, and the planning data will be used for those not available, and the planning forecast data will be used for all the years from 2019 to 2020.

EXAMPLE ANALYSIS AND DISCUSSION

The model combining the entropy method and set pair analysis was used to evaluate water security within Ningxia. The former method provided the weight of each indicator whereas the latter determined the level of safety. The method considered the fuzziness of the equal-level boundary and the differences between the weights of each evaluation index. The results of the method avoided subjective randomness.

To construct a set of water security indices $H_{2006\sim2013,2018\sim2020}(A_{19},B_5)$ for water resources in Ningxia, the single degree $\mu_{A_i\sim B_1}$ of contact for all indicators was first calculated, thereby providing the weight of each indicator. The final total was the calculated degree of contact $\mu_{2006\sim2013,2018\sim2020}(A_{19},B_5)$, within the interval [−1, 1] which was further divided into five intervals (Huawei *et al.* 2015; Wenjuan *et al.* 2018; Yongan & Wensheng 2018), and the tenth to 15th columns of Table 3 correspond to the five levels of the water security evaluation from high to low and from left to right. First, the set results were compared with the corresponding values of each level to obtain the classification of water security status for eight consecutive years. Water security status μ had a positive relationship with the connection value and water security status. Finally, the water security status was evaluated and analyzed and development trends were identified. At the same time, the confidence criterion was calculated, and the confidence attribute h_k of each year was determined to obtain the level of water security.

Determination and analysis of entropy weight

The weights of the indicator layer and criterion layers were determined based on the entropy weight method, the results of which are shown in Table 3.

Evaluation of index layer

According to the evaluation results of the index layer (Table 3), the influence of the 19 indicators in the index layer on the security of water resources in Ningxia from largest to smallest was: C19 > C1 = C5 > C18 > C14 > C10 > C13 > C3 = C4 > C8 > C7 > C15 > C16 > C12 > C2 > C11 > C17 > C6 > C9.

The top three indicators represent the basic quantity and quality conditions to ensure water security, thereby indicating that the conditions of water resources in this region are an important constraint on water security. Excluding the top three indicators, water security constraints, water resources development and utilization rate and other water resource conditions, the ranks of indices 4–10 were in the following order: C18 > C14 > C10 > C14 > C5 > C8 > C7, where one index falls into the category of coordination with the aquatic environment, three indices fall within coordination capabilities, two fall within economic coordination capabilities and one falls within water resources conditions. Five of the seven indicators account for the coordination between socio-economic factors and the environment, which fully demonstrates the important position and role of coordination between society, economy and aquatic environment under the basic condition of 'water quantity and quality'. The effects of indices 6 and 4 within economic and social coordination and the consequent impacts on water resources security are far greater than those of indices 7, 9 and rate of uptake of urban water-saving appliances, which indicates that for areas such as Ningxia where agricultural water consumption is extremely high, the weights of indices 6 and 4 on water security are much greater than that of 7. Compared with urban water resources, the regulation of agricultural water resources utilization provides a greater benefit to water security, and is therefore a key point on which to focus in Ningxia.

Evaluation of criteria layer

According to the calculation result of Equations (5)–(9), the water resources condition criteria layer has a weight of 0.313, a social coordination capacity of 0.186, an economic coordination capacity of 0.187 and a water environment coordination capacity of 0.313. Since water resources are scarce and the water environment is poor, this indicates that the basic conditions of water resources and water environment coordination capacity remain important in restricting water security in Ningxia. Both contribute >60% of the binding force, and the influence of social coordination capacity and economic coordination capacity accounts for approximately 37%. However, as shown by changes in the basic data (Figure 1), the relative change of the basic conditions of water resources during 2006–2013 was very small, except for a change in the rate of C4 to 45%. For example, the rate of change of C5 was the largest, but only 5.8%. C1 was not affected by the various factors, but the five indicators of social coordination capacity, C6–C10, showed rates of change of 17.2%, 25.7%, 63.2%, 56.3% and 92.8%, respectively. The rates of change for C11–C14 were 70%, 74%, 83.9% and 6.0%, respectively. The rates of change of water environment coordination capacity of C15–C19 were 56.5%, 36.7%, 40%, 21% and 5.93%, respectively. The rate of change of each indicator is shown in Figure 1. From the perspective of relative change rate, under the condition of basic conditions of water resources remaining mostly unchanged, the obvious changes in social, economic and aquatic environment coordination ability are an important factor driving improved regional water resources security, particularly related to water-saving measures. Important indices include the efficiency indices C8–C10, benefit indices C12–C13 and environmental protection investment and efficiency indicators C11 and C15–C18, while reclaimed water reuse is also an important driving factor. After 2006, investments in the social, economic and aquatic environment of Ningxia greatly improved. Soon after, the water-saving level, water consumption level, management level, and economic input level of Ningxia improved rapidly. The changes in regulation efficiency indicators reflect the significant contribution of socio-economic factors, input of resources to the aquatic environment and regulation of the impact of regional water resources security over the past eight years. Under unfavorable conditions of natural endowment of water resources, it is necessary and effective to strengthen and improve the water security status of the region through social and economic regulation and management in line with the shortage of water resources.

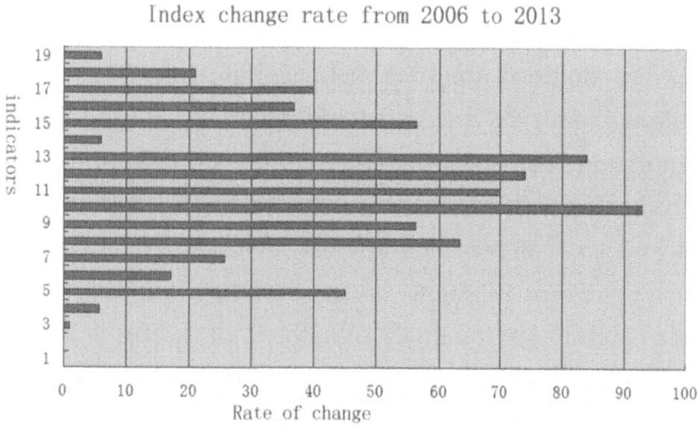

Figure 1 | Rates of change of each indicator during 2006–2013.

Evaluation of the contribution degree in different years

To further clarify the evaluation results, the contribution of the indicators of the eight different levels of the year to water security was obtained and the rankings of the top six are shown in Table 4. It is evident that the impact of different indicators on regional water security differs for different levels.

From 2006 to 2013, the impact of different indicators in different levels on water resources security differed, and with the gradual development of economic and social regulation and control, the constraints of water shortages on water security declined. (1) During the initial stage of water-saving interventions in 2006, the basic conditions of the water resources of the region remained the main

Table 4 | Sorting the contribution of indicators in different years to water resources security

Years	Top six indicators of contribution	Years	Top six indicators of contribution
2006	C7 > C4 > C1 > C2 > C5	2011	C18 > C13 > C15 > C14 > C12 > C3
2007	C5 > C18 > C7 > C9 > C15 > C1	2012	C18 > C13 > C15 > C12 > C8 > C3
2008	C18 > C5 > C7 > C13 > C15 > C8	2013	C18 > C5 > C13 > C15 > C14 > C12
2009	C18 > C15 > C13 > C7 > C8 > C12	2018	C1 > C18 > C10 > C14 > C15 > C4
2010	C18 > C15 > C13 > C8 > C3 > C12	2019	C1 > C18 > C10 > C14 > C15 > C4
		2020	C1 > C18 > C14 > C10 > C15 > C4

constraint. Water reuse had just started, which alleviated water shortages. (2) The implementation of water-saving interventions during 2007–2008 was fully implemented, which had a strong impact on overall water security, water management, wastewater reuse, urban water saving and reducing urban water losses. (3) On the basis of improved water quality and water loss reduction during 2009–2011, C13 and C12 became increasingly important for water security. (4) From 2012 to 2013, the rapid development in urban water-saving greatly improved water security; however, the large proportion of agricultural water use and the inefficient use of water became the constraints on the efficiency of alternative industrial regulation.

In general, in regards to water shortages, an increase in the regulation of the socio-economic environment improved water quality and the aquatic environment, water consumption efficiency increased and the degree of C4 and C14 had a large impact on water security during the study period. The enhancement of economic and social regulation and control weakened the impact of water shortages on water security. For the indicators with small annual contributions, such as C10 and C11, the contribution rate in each year was the lowest, indicating that this region with an underdeveloped economy has a large proportion of agricultural water and low degree of water-saving irrigation. It is also evident that all those indicators remain the important constraints affecting water security. Economic development determines the input of water security.

(5) Data for 2018–2020 were used for planning. The construction of a water-saving society is a whole society water-saving strategy to mobilize various economic and social resources and is carried out with high-intensity investment. Ningxia is one of the pilot projects. In this paper, the pilot construction period of the water-saving society with greater economic and social regulation intensity from 2006 to 2013 is selected as the research historical data, and the planning data of water-saving society construction from 2018 to 2020 are used for comparison, so as to explore the impact of different regulation intensities on water security and its change trend. It is evident that during the planning year, according to the current regulation and intensity, within the ranking of the degree of contribution to regional water security, the per capita water resources and water resources utilization rate reappear in the top six, and the discharge rate of industrial wastewater reaches the standard from the first place during 2008–2013 to the first place in the ranking. Secondly, C10 and C11 are included in the ranking, indicating that under the planning and control measures, the efficiency of industrial regulation and control is improved and is faster, the efficiency and ability for agricultural regulation and control does not match the level of social development, and the factors contributing to a greater degree to regional water insecurity gradually shift from industry to agriculture. The capacity of agricultural, economic and social regulation and control during the planning year cannot meet the social–economic demand for quantity and quality of water resources under resource constraints. The capacity of regulation and control has declined, and the binding force of water resources has increased. The basic conditions of water resources have once again become a constraint on water security.

(6) From the ranking of the indicators during 2006–2020, changes to the driving forces of regional water security are evident: during initial implementing of water-saving interventions, the population and basic conditions of water resources are the main factors (C7 > C4 > C1 > C2 > C5 in 2006), whereas during 2008–2013, the driving forces are concentrated in C18, C15, C13, C12, C3 and C14 and other ecological indicators. The indicators driving the capacity for environmental and economic coordination for the planning year for 2008–2013 are C18, C15 and C14. The per capita water resources amount C1 and water

resources development and utilization index C5 re-enter the main indicators, ranking first and sixth, respectively, and the rankings of the proportion of agricultural water-saving irrigation area C10 and the proportion of agricultural water use C14 have upward trends, indicating that for this region, along with the enhancement of regulation of socio-economic and ecological environment, congenital restraint of water resources has been improved to a certain extent, and water security has been enhanced. However, during the planning level year, shortages of water have once again become a major factor restricting water security. The contribution of the ability for socio-economic regulation on water security during the planning year cannot match the demand for water resources during the planning stage. According to current regulation intensity, the influence of water resources conditions on regional water resources security will again become the main constraining factor.

Evaluation of safety level

Table 3 shows the results of set-pair confidence analyses. The results of the two methods were consistent. The set-pair connection method showed the status of water resources in Ningxia during 2006 to be unsafe (4), of basic safety (4–3 during 2007–2013 and insecure during 2006–2008), and then of regional basic security. Water security within Ningxia is improving, although it remains at a basic security state; however, the degree of connection and confidence indicate that water security is moving towards a secure state. The grade of water resources security for Ningxia from 2006 to 2008 is 4 (insecurity and basic security), possibly because water resources remain insufficient, and C8, C10, C9 and C14 under the coordination ability of society, economy and the ecological environment. The standard is in an extremely unsafe state, which directly affects the water security of Ningxia in 2006, and the social, economic and ecological environment has not played a coordinating role in water resources, which impacts sustainable development. Since 2006, the effect of water-saving interventions in Ningxia has been reflected. With water resources remaining unchanged, other indicators have improved greatly; therefore, water security in Ningxia has gradually improved.

Table 5 shows the degree of connection, result of the confidence calculation and the corresponding rating of Ningxia's water security. According to the calculation results, the changes to water security of the water resources of Ningxia are drawn on the linkage line chart, as shown in Figure 2. From 2006 to 2013, water security (Figure 2) tended to move toward level 1. The probability of moving to levels 2 and level 3 decreased with time. The probability of moving toward level 4 increased slightly in 2011 compared with 2010. Water security was in a declining state for the remaining time, and the possibility of tending to level 5 decreased gradually. By examining the basic water resources data in 2011, it was found that rainfall and per

Table 5 | The calculated and rating results of the water security evaluation set of Ningxia in 2006–2013 and planning level year

years	f_1	f_2	f_3	f_4	f_5	μ Total contact	μ Determine security level	h_k	h_k Security level
2006	0	0.113	0.358	0.156	0.374	−0.3955	IV	0.627	IV
2007	0.074	0.129	0.311	0.116	0.370	−0.2895	IV	0.514	III
2008	0.125	0.137	0.259	0.122	0.358	−0.2255	IV	0.521	III
2009	0.288	0.080	0.158	0.124	0.350	−0.084	III	0.526	III
2010	0.331	0.078	0.140	0.129	0.322	−0.0165	III	0.549	III
2011	0.342	0.084	0.123	0.172	0.280	0.018	III	0.549	III
2012	0.382	0.054	0.172	0.122	0.270	0.078	III	0.608	III
2013	0.403	0.059	0.172	0.142	0.224	0.1375	III	0.625	III
2018	0.468	0.029	0.188	0.143	0.171	0.24	III	0.998	III
2019	0.469	0.038	0.186	0.136	0.171	0.249	II	0.999	II
2020	0.469	0.046	0.187	0.127	0.171	0.2575	II	0.999	II

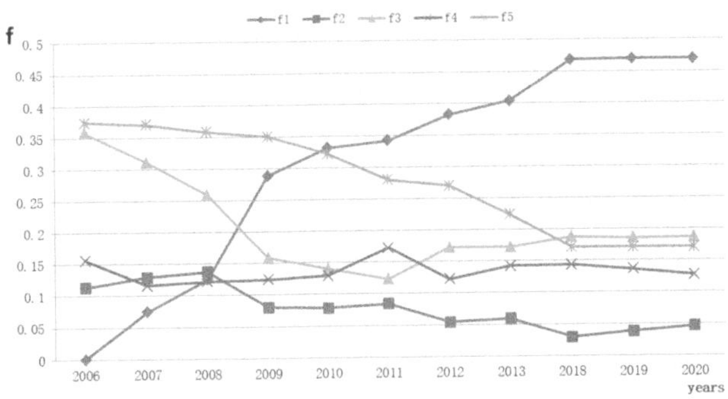

Figure 2 | Ningxia 2006–2020 water resources safety assessment set.

capita water resources in 2011 were lower than those in 2010 due to climate conditions, which may lead to fluctuations in the evaluation results. This illustrates that the possibility of extreme water insecurity in Ningxia is gradually decreasing and the possibility of regional security is increasing.

CONCLUSIONS AND RECOMMENDATIONS

The evaluation of water security based on social and economic regulation was discussed, and a water security evaluation system was established. Taking Ningxia as an example, we choose indicators reflecting the efficiencies of economic, social and aquatic environment regulation to represent the status of water security. The present study reflected the dynamic process of social economic regulation and control on water security using the entropy weight–set pair model. The data were evaluated for eight consecutive years. The results showed that under conditions of rigid constraints on water resources, economic and social control measures, such implementation of water-saving interventions, the most stringent water resources management system and investment within the aquatic environment can have obvious benefits for water security in the arid areas of Ningxia.

The assessment of water security based on the efficiency of social and economic regulation is suitable for arid regions and can reflect the main contradictions and directions for improving water security in arid regions.

(1) From 2006 to 2013 and 2018–2020, regional water resources security has gradually increased from insecurity to basic security, as well as security in the planning period, the impact of different indicators at different levels on water resources security differed, and with the gradual development of economic and social regulation and control, the constraints of water shortages on water security have declined.

(2) According to the evaluation results of the criteria layer, the basic conditions of water resources and the coordination ability of the water environment are still the important factors restricting water security in Ningxia. From the perspective of relative change rate, under the condition of basic conditions of water resources remaining mostly unchanged, the obvious changes in social, economic and aquatic environment coordination ability are an important factor driving improved regional water resources security, particularly related to water-saving measures.

(3) During the pilot construction of the water-saving society, at the initial stage of regulation, the contribution degree of water resources conservation and supplementary indicators and measures was higher, and the social and economic control measures in the middle and late stages of regulation gradually played a greater role; however, in the planning year, the factors affecting regional water resources security changed to some extent, and the shortage of water resources once again became a constraint factor, indicating the economic and social regulation power of the planning year. The degree of water resources and the contradiction between supply and demand of water resources do not match, which cannot effectively alleviate the regional water resources security situation.

(4) Among the factors affecting water resources security, the benefit index reflecting the coordinated development of society, economy and the water environment is an important index to promote regional water security. However, due to the slow development of regional regulation and control capacity, agricultural water use efficiency has not kept up. Strengthening social and economic coordination can improve water security and alleviate the impact of water shortage. However, if the capacity of water safety supervision cannot keep up with the needs of social and economic development,

the role of social and economic regulation in water security will be weakened.

In the future, the region should further strengthen the ability and means of social and economic regulation and control. The measures that can be taken are: continue to increase the area of agricultural water-saving irrigation, reduce the proportion of agricultural water use, improve the efficiency of agricultural water use, improve the rate of industrial sewage discharge reaching the standard, further reduce the water consumption of 10,000 yuan industrial added value, and improve the proportion of recycled water. Through these measures to further reduce agricultural water consumption, improve water efficiency, through agricultural water-saving to achieve ecological water, domestic water, and industrial water security support, and improve the overall security of regional water resources.

ACKNOWLEDGEMENTS

This study was supported by the Major Innovation Project for Building First-class Universities in China's Western Region (ZKZD2017002), Key Laboratory for Restoration and Reconstruction of Degraded Ecosystem in North-Western China of Ministry of Education, the Key Research and Development Program of Ningxia Hui Autonomous Region (2018BEG03008), the Project of First-class Discipline Construction (Domestic First-class Construction Discipline) in Ningxia Colleges and Universities (NXYLXK2017A03), the National Natural Science Foundation of China (51269022).

DATA AVAILABILITY STATEMENT

All relevant data are included in the paper or its Supplementary Information.

REFERENCES

Handmer, J. W., Lustig, T. L. & Smith, D. I. 1986 Assessing intangible flood damages for evaluating urban floodplain management options. *Hydrology and Water Resources Symposium* **52** (5), 1567–1570.

Huaiwen, C., Yuwen, L. & Zhongmin, X. 2011 The theory and empirical evidence of social adaptability of water shortage – taking the Heihe River Basin as an example. *Journal of Ecology* **31** (5), 1430–1439 (in Chinese).

Huawei, Z., Rui, Z., Zhan, M. & Youzhou, L. 2015 Cultivated land ecological safety diagnosis based on PSR model and set pair analysis. *China Land Science* **29** (12), 42–50 (in Chinese).

Jun, X. & Wei, S. 2016 Research and prospects of China's water safety in changing environments. *Journal of Hydraulic Engineering* **47** (3), 292–302 (in Chinese).

Jun, X., Chunxi, L. & Guoyu, R. 2011 Opportunities and challenges faced by climate change impacts on China's water resources. *Advances in Earth Science* **26** (1), 1–12 (in Chinese).

Jun, S., Yingwei, O., Jinfeng, C. & Wei, G. 2016 Water resources security assessment of the Yangtze River Basin based on water poverty index. *Resources and Environment in the Yangtze Basin* **25** (6), 889–895 (in Chinese).

Junyuan, S., Fengping, W. & Qianwen, Y. 2016 Comprehensive evaluation of the most stringent water safety based on fuzzy set pair analysis. *Journal of Water Resources and Water Engineering* **27** (2), 92–97 (in Chinese).

Li, G. & Songhong, Y. 2014 Urban water security evaluation based on set pair analysis method. *Journal of Lanzhou Jiaotong University* **33** (1), 149–153 (in Chinese).

Liping, L. & Deshan, T. 2014 Analysis of water resources shortage and social adaptability evaluation and coupling coordination relationship. *Arid Land Resources and Environment* **28** (6), 13–19 (in Chinese).

Min, L., Zhan-yu, Z. & Yuezhen, S. 2006 Application of set pair analysis method in water safety evaluation. *Journal of Hohai University (Natural Science Edition)* **34** (5), 505–509 (in Chinese).

Ohlsson, L. 1998 Water and social resource scarcity – an issue paper commissioned by FAO/AGLW. In: *2nd FAO E-Mail Conference on Managing Water Scarcity, WATSCAR 2.*

Ohlsson, L. 2000 Water conflicts and social resource scarcity. *Physics and Chemistry of the Earth, Part B: Hydrology, Oceans and Atmosphere* **25** (3), 213–220.

Page, G. W. 2001 Planning implications of water supply decisions. *Planning Practice & Research* **16** (3–4), 281–292.

Qiting, Z. 2017 Water resources adaptive utilization theory and its application prospect in water control practices. *South-to-North Water Transfer and Water Science and Technology* **15** (1), 18–24 (in Chinese).

Wen, D., Jinfeng, W., Shibin, M. & Chanlong, D. 2014 Comprehensive evaluation of water resources security based on set pair analysis method. *Water Science and Engineering Technology* **4**, 38–42 (in Chinese).

Wenjuan, M., Chunling, P., Qianli, C., Ling, J., Lili, S., Feixiang, M. & Xiangxin, L. 2018 Health evaluation of land use system in Urumqi based on entropy weight–set pair analysis method. *Chinese Journal of Agricultural Resources and Regional Planning* **39** (2), 29–36 (in Chinese).

Xia, Z., Bing, H. & Dan, W. 2018 Water resources security evaluation and regional differences in the process of new urbanization. *Journal of North China University of Water Resources and Hydroelectric Engineering* **34** (4), 14–19 (in Chinese).

Yiping, W., Lianglin, W. & Shike, Z. 2013 Comprehensive evaluation of water resources security in Hechi city. *Water Science and Engineering Technology* **2013** (4), 6–9 (in Chinese).

Yirong, S. & Jiancang, J. 2014 Research on water security fuzzy matter element evaluation model based on entropy weight and TOPSIS method and its application. *Systems Engineering* **32** (7), 143–147 (in Chinese).

Yong-an, L. & Wen-sheng, W. 2018 Application of set pair analysis method in the optimization of urban flood control standards. *Journal of North China University of Hydraulic Engineering (Natural Science Edition)* **39** (1), 77–81 (in Chinese).

Zhengwei, P., Yixing, Z. & Yanhua, W. 2016 Set pair analysis and variable fuzzy set method for vulnerability assessment of regional water resources system. *China Population, Resources and Environment* **26** (11), 198–203 (in Chinese).

Zhifei, L. 2017 Water resources security and implementation of the 'one belt, one road'. *Strategy* **17** (4), 45–54 (in Chinese).

First received 15 June 2020; accepted in revised form 15 October 2020. Available online 28 October 2020

Research on ecological environment impact assessment based on PSR and cloud theory in Dari county, source of the Yellow River

Wang Yongqiang, Wu Zhijun, Yan Bo, Li Kai and Huang Feng

ABSTRACT

In order to reasonably evaluate the ecological environment of Dari County in the source region of the Yellow River, the characteristics and actual conditions of Dari County's ecological environment are taken into account, and the principles of representativeness, scientificity, operability and systematicness of index are followed. An evaluation index system is established for Dari County based on the Pressure-State-Response (PSR) conceptual framework. Combining the Analytic Hierarchy Process (AHP) with the cloud model, an ecological environment evaluation model for Dari County is established, and the ecological environment of Dari County is quantitatively evaluated. The model organically combines the ambiguity and randomness of the uncertainty concept. It not only describes the ambiguity of the evaluation level with a membership function, but also considers the randomness of the membership itself using the concept of superentropy, which is more in line with the actual evaluation object. The results indicate that the ecological situation of Dari County is in a 'general' state that should be urgently protected for sustainable development, and land degradation is the most important factor affecting the ecological environment of Dari County.

Key words | assessment, cloud model, Dari County, ecological environmental impact, PSR framework

Wang Yongqiang
Wu Zhijun
Yan Bo (corresponding author)
Li Kai
Water Resources Department (International River Management Department),
Changjiang River Scientific Research Institute,
Wuhan 430010,
China
and
Hubei Provincial Key Laboratory of Basin Water Resources and Ecological Environmental Sciences,
Changjiang River Scientific Research Institute,
Wuhan 430010,
China
E-mail: *byanhhu@163.com*

Wu Zhijun
Huang Feng
College of Hydrology and Water Resources,
Hohai University,
Nanjing 210098,
China

HIGHLIGHTS

- A model method combining AHP and cloud model.
- The ecological environment of Dari County is in a general state.
- The evaluation index system of Dari County based on PSR framework is proposed.
- Land degradation is an important factor affecting the ecology of Dari County.
- The evaluation model can take into account both ambiguity and randomness.

INTRODUCTION

The source region of the Yellow River is located in southwest Qinghai Province, the hinterland of the Qinghai-Tibet Plateau. It is an important water conservation area and ecological barrier of the Yellow River, as well as one of the most special and fragile areas in China. With the influence of climate change and human activity, the ecological environment in parts of the source Yellow River areas have suffered damage in recent years. To protect the ecological environment, it is necessary to know and quantitatively evaluate the current state of the regional ecological environment. The difficulty is how to select the corresponding evaluation index system according to the regional characteristics and

doi: 10.2166/ws.2020.286

make a reasonable quantitative analysis. In view of this, assessing the state of the ecological environment in the river source area is of great significance for its protection and restoration. This paper selects the 'Pressure-State-Response (PSR)' model as the index system construction model. At present, domestic research scholars have widely used this model in the fields of water environment safety assessment, urban planning environmental impact assessment, and wetland ecosystem health assessment. For example, Li Lin's study on the evaluation of urban lake water environment safety divides the evaluation indicators of urban lake water environment safety into three levels: pressure-state-response. The pressure layer selects the most threatening indicators for lake water environment safety, and the state layer selects the representative lake State indicators, the response layer selects indicators that reflect the measures taken in lake management. Xu Youpeng's wetland ecological health assessment based on the 'stress-state-response' model and fractal theory: stress reflects the impact and stress of human activities on the wetland ecosystem, reflects the intensity and change trend of resource utilization; state reflects the wetland ecological environment, the change of elements reflects the ultimate goal of the environmental protection policy; the response reflects the measures taken by the society or individuals to stop, mitigate, prevent or restore the changes of the wetland ecosystem. The PSR model framework provides an evaluation idea, emphasizing the causal logic of the ecological environment impact process, and the need to analyze the change process of the evaluation object. The model is simple and easy to use and is widely used. Foreign scholars have applied the PSR model to environmental sustainability, regional sustainable development, and ecological security Lin Zhen's (Zhen *et al.* 2009) Comparison of sustainability issues in two sensitive areas of China uses the PSR model to select the indicators of two sensitive agricultural areas in China to assess the sustainability of the two areas. Yang Pei's (Yang *et al.* 2011) Ecological Risk Assessment of the Shenzhen River-Bay Watershed builds a pressure-effect-social response (PESR) ecological risk Evaluation System. The fuzzy comprehensive evaluation method (FCE) is used to calculate the risk index of each subsystem. At present, selection of specific indicators for ecological environment evaluation needs to consider different regional characteristics. For example, He Sanwei (Sanwei *et al.* 2011) reflects the ecological environment of agricultural land from the three aspects of water, soil, and climate based on PSR and cloud theory, considering fertilizer, irrigation condition and the degree of mechanization, and then a PSR-based agricultural land ecological environment evaluation index system is constructed. Zhang Zezhong (Zezhong *et al.* 2014) comprehensively assesses the current status and changes of ecological impacts of the North Canal from the perspective of river hydrological conditions, water environmental conditions, habitat quality and biological conditions. Compared with the ecological environment studies on watershed and agricultural land, there are fewer on the source region of the Yellow River (Zhongzheng *et al.* 2003). Wei Jing's evaluation of the ecological environment vulnerability in the Three Rivers source area (Jing *et al.* 2015) selects meteorological factors, topographic factors, surface factors, vegetation net primary productivity (NPP), vegetation coverage factors and indicators of human and economic activities for comprehensive evaluation based on the definition of ecological environment vulnerability and ecological sensitivity-ecological resilience-ecological pressure (SRP) model. Wu Qing (Qing & Yanli 2002) systematically analysed the causes of water resources fragility in the source Yellow River region, such as the uneven spatial and temporal distribution of climate and rainfall, grassland degradation, land desertification, and human activities, and its ecological impacts. Based on the latest research results, this paper attempts to put forward the PSR framework eco-environmental impact assessment index system in Dari County, the source region of the Yellow River.

In the existing researches, the domestic ecological environment assessment of river basins, plateau regions or eco-environment impacts of projects are mainly focused at home, while ecological impact assessments from certain perspective are concerned abroad (Bartsev *et al.* 2012; Carlos *et al.* 2020). But so far, related to regional ecological environment impact assessment, models are not clear and methods are close at home and abroad. The evaluation methods are mostly fuzzy comprehensive methods (Jun *et al.* 2013), Back Propagation (BP) neural network method, system dynamics (Hjorth & Bagheri 2005) and other related methods. The BP neural network method results in problems of inconsistency in the actual process.

The fuzzy comprehensive method divides the fuzzy classification boundary of indexes, but the calculated membership degree is easily disturbed by subjective conditions. The system dynamics method is to build a model around the problem; results will vary greatly because of different models for same problem. In summary, the existing quantitative evaluation methods and models can obtain regional ecological environment impact level, but cannot explain the level conversion status, because they ignored the ambiguity of the quantitative index description and the randomness of the index value determination and rating division. Academician Li (1998) first proposed the cloud model, which can take both ambiguity and randomness of evaluation into account, and has been widely used in multi-attribute decision-making, analysis and evaluation. For example, Zhang Yang's (Yang *et al.* 2013) assessment of land resource ecological security in Hubei Province is based on the normal cloud model. This study introduces the normal cloud model into the assessment of regional land resource ecological security. Taking Hubei Province as an example, a quantitative measurement of the regional land resource ecological security status is made. In Gao Mingmei's (Mingmei *et al.* 2015) application of the normal cloud model in the evaluation of land ecological security in Wanjiang area, the concept portrays the randomness of the membership degree itself, and obtains the land ecological security level of Wanjiang area.

The source region of the Yellow River studied in this paper has the characteristics of ecological fragility and has ambiguity and randomness from the perspective of mathematical theory. Therefore, we introduce the PSR indicator system framework and cloud model theory. The 'Pressure-State-Response' evaluation model is different from the traditional ecological evaluation framework. Organizing and classifying ecological indicators based on the interaction and impact of the ecosystem has a strong systematic and distinct ecological significance. The cloud model is used to convert between qualitative concepts and quantitative values, and is an important method in the evaluation of uncertain artificial intelligence. In this paper, Dari County, which is in the source region of the Yellow River, is taken as the study area. Considering its local ecological environment characteristics and actual conditions, and following the principles of indicator system construction such as representativeness, scientificity, operability, and systematicness, the 'Pressure-State-Response' indicator conceptual model is established. Based on the actual conditions of the Yellow River source which is a water conservation ecological function area, the ecological environment impact indicator system of Dari County was constructed, and the orderly levels of each indicator in the evaluation target and the objective judgment of expert opinions were analyzed by the analytic hierarchy process (Ying *et al.* 2007). Combine them effectively and use mathematical methods to determine weights. Then the cloud model is used to quantitatively evaluate the ecological environment impact of Dari County, for objectively giving its environment status.

STUDY AREA AND DATA SOURCE

Study area

Dari County is located in the Sanjiangyuan District (Figure 1). The source area of the river is located in the southwest of Qinghai Province in the hinterland of the Qinghai-Tibet Plateau and is the source of the Yangtze, Yellow and Lancang Rivers. It is an ecological barrier for the ecological environment and regional sustainable development of the middle and lower reaches of China's rivers and Southeast Asian countries. Dari County is the core area of the Sanjiangyuan, and its ecological location is very important. It is of great significance for ensuring the safety of the ecological environment of the Sanjiangyuan and promoting the development of the regional ecological environment (He *et al.* 2018). In recent years, regional climate change and intensified human activities have caused a decrease in precipitation, which has reduced water conservation functions, reduced water production, and reduced forest and grass coverage. Regional socio-economic development has had a serious impact (details in Table 1).

Data source

The basic data used in this paper is collected from China National Data Network (http://data.stats.gov.cn/), Chinese Academy of Sciences Resource and Environmental Science Data Center (http://www.resdc.cn/), Qinghai Meteorological Bureau Climate Data Center, field survey, annual statistics of various indicators.

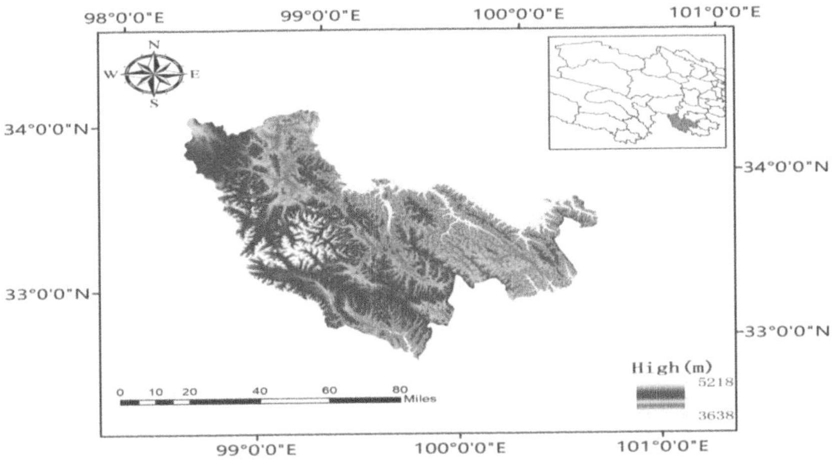

Figure 1 | Geographical location of the research area in Dari County, Qinghai Province.

Table 1 | Characteristics of ecological and environmental in Dari County

Name	Data	Name	Data
Annual rainfall	570 mm	Aridity index	1.96
Annual average temperature	−1.3 °C	Natural grassland area	$140 \times 10^4 hm^2$
Annual evaporation	1,120 mm	Utilizable grassland area	$111 \times 10^4 hm^2$
Wind speed	34 m/s	Deteriorated grassland area	$78 \times 10^4 hm^2$
Total land area	$148 \times 10^4 hm^2$	'Black Soil Type' deteriorated grassland area	$57 \times 10^4 hm^2$

REGIONAL ECOLOGICAL EVALUATION INDEX SYSTEM BASED ON PSR

PSR

The PSR model is an indicator selection framework based on the interconnections between indicators. It grasps the development process of the ecological environment impact by analyzing status changes. According to the environment of different areas, the corresponding indicators are selected to constitute the index layer, so as to focus on the ecological environment of Dari County. Then, an ecological environment impact assessment index system based on PSR and a target framework suitable for Dari County were constructed.

According to the PSR model framework, pressure index refers to the interference and pressure factors from external conditions in the evaluation index system constructed, state index is used to reflect the ecological environment, and the response category refers to the impact of the ecosystem under the current ecological environment pressure (Long *et al.* 2012). Interconnect the three aspects of pressure-state-response to build the framework of the ecological environment impact assessment system, and select specific indicators to form a three-level indicator system.

Evaluation index system for Dari County

The construction of an assessment index system is the basis for conducting an ecological environment impact assessment. But considering that different regions have different characteristics, a specific research indicator system and evaluation model need to be carried out for the study area, based on its ecological environment characteristics and actual conditions. Details of the eco-environmental impact assessment index system of Dari County are in Table 2. Apart from the influence of water resources, mainly climate factors, grassland degradation and land desertification are major factors that cause ecological and environmental

Table 2 | Index system of ecological environment impact assessment in Dari County

Target layer	Criterion layer	Indicator layer	Indicator meaning
Ecological Environmental Impact Assessment	Pressure	Annual precipitation P1	The impact of rainfall on vegetation growth
		Drying index P2	The pressure of climate dryness
	State	Grass coverage S1	Represents a green sustainable state, the foundation for water renewal
		Soil salinity S2	Indicates the local soil salinity status
		Soil water content S3	State of soil water storage capacity
	Response	Soil erosion area ratio R1	Response to grassland degradation
		Land degradation rate R2	Response to the formation of 'black soil beach"
		Soil erosion modulus R3	Responds to soil holding capacity

impacts in Dari County. In the construction of the index system, the climate and the degree of dryness reflect the disturbance and pressure factors of the water resources, which is an important factor in environmental changes. The state of soil salinity, soil moisture content, and grassland coverage are selected to reflect the ecological environment condition of Dari County. Degradation of grassland weakened the regulation of vegetation on land runoff. The response of the ecological environment refers to the impact of the ecosystem under the current ecological environment pressure, which is mainly represented by soil erosion and land degradation.

Grading standards

There are regional differences in the standards of the ecological environment index levels. This study takes into account the actual situation of Dari County in the source region of the Yellow River, with reference to the research results of ecological index thresholds at home and abroad, ecological environment evaluation standards, standards issued by local governments, and planning objectives, etc. (Jinrui *et al.* 2003; Jian *et al.* 2006) The eco-environment evaluation level of Dari County in the source region of the Yellow River is divided into four levels: poor, fair, good and excellent. Finally, combined with the opinions of experts, the classification standards for regional ecological environment evaluation indicators are obtained. The specific ecological environment classification standards are shown in Table 3.

CLOUD MODEL EVALUATION METHOD

Cloud model

The cloud model (Lujun & Yaping 2000) is an uncertainty model proposed by Academician Li Deyi that can be used to represent both a certain qualitative concept and its quantitative concept. It shows an advantage in the analysis of the randomness and ambiguity of the thing itself for unified analysis. The model is a new mathematical model, which is characterized by a combination of normal distribution and

Table 3 | Classification standard for ecological environment impact assessment in Dari County

Target layer	Criterion layer	Indicator layer	Evaluation standard			
			I (Poor)	II (General)	III (Good)	IV (Better)
Ecological Environment Impact Assessment	Pressure	Annual precipitation P1	<1,000	1,000–1,300	1,300–1,500	>1,500
		Drying index P2	>4	4.0–1.2	1.2–1	<1
	State	Grass coverage S1	<15	15–20	20–25	>25
		Soil salinity S2	>5	3–5	2–3	<2
		Soil water content S3	<5	5–12	12–15	15–20
	Response	Soil erosion area ratio R1	>35	20–35	5–20	<5
		Land degradation rate R2	>10	6–10	4–6	<4
		Soil erosion modulus R3	>5,000	2,500–5,000	1,000–2,500	<1,000

membership function and has a wide range of universality. The three digital characteristics of the cloud model are expected value (*E*x), entropy (*E*n), and superentropy (*H*e) (Yang *et al.* 2013). Ex represents the grade point value of the ecological environment impact assessment of Dari County, while *E*n divides the index level. The degree of uncertainty and dispersion of each index is expressed by He, which indicates the degree of correlation between randomness and ambiguity within indicators. The three characteristic value parameters (*E*x, *E*n, *H*e) of each grade cloud are determined by the upper and lower boundary values of each grade of the evaluation index corresponding to the grade, which can be obtained by the corresponding characteristic value parameter value formula. There is data x_{ij}, where i is the evaluation index, and j is the evaluation level corresponding to the data x; then there are x_{ij}^1 and x_{ij}^2 for the upper and lower boundary of the data x_{ij}. Model representation can obtain an i * j cloud model. Since the intermediate value of each level is the qualitative concept that can best represent that level, the expected value is expressed as:

$$Ex_{ij} = \frac{(x_{ij}^1 + x_{ij}^2)}{2} \tag{1}$$

As the boundary of each level, x_{ij} belongs to both the previous level and the next level, so the boundary value is equal to the membership degree of the upper and lower levels, thereby obtaining the entropy value:

$$En_{ij} = \frac{(x_{ij}^1 - x_{ij}^2)}{6} \tag{2}$$

The size of *H*e is generally obtained based on the entropy value and experience, which mainly reflects the thickness of the cloud layer, which is 0.01 here:

$$He = T \tag{3}$$

Ecological environment impact assessment model based on cloud model

In the process of quantitatively measuring the regional ecological environment, the quantitative description of its evaluation indicators has both ambiguity and randomness. If only the ambiguity or randomness of the objects in the regional ecological environment evaluation is considered, then an accurate evaluation result is expected. The purpose is difficult to achieve, so it is necessary to introduce a model method that can take into account both ambiguity and randomness. This study builds on the normal cloud model theory and builds a comprehensive evaluation model based on the normal cloud. The construction of the evaluation model is mainly divided into the following parts: (Figure 2).

- Establish an index system (X1, X2, X3, …, Xn) of the ecological environment impact assessment, and formulate an evaluation level standard V = {V1, V2, …, Vm}.
- Evaluate each index in the system and establish a fuzzy matrix R. The element rij in R represents the degree of membership of the i-th element in the system.

$$r_{ij} = (Ex_{ij}, En_{ij}, He), \qquad R = \begin{pmatrix} r_{11} & \cdots & r_{1j} \\ \vdots & \ddots & \vdots \\ r_{i1} & \cdots & r_{rj} \end{pmatrix}$$

- Take the quantitative value rij of each index as the input for the cloud generator to get a set of cloud drops (a, y). Execute the cloud generator multiple times, and determine the cloud membership matrix $Z = (Z_{ij})_{n*m}$. Cloud generator algorithm: First set *E*x as the expected value and *H*e as the standard deviation to generate a

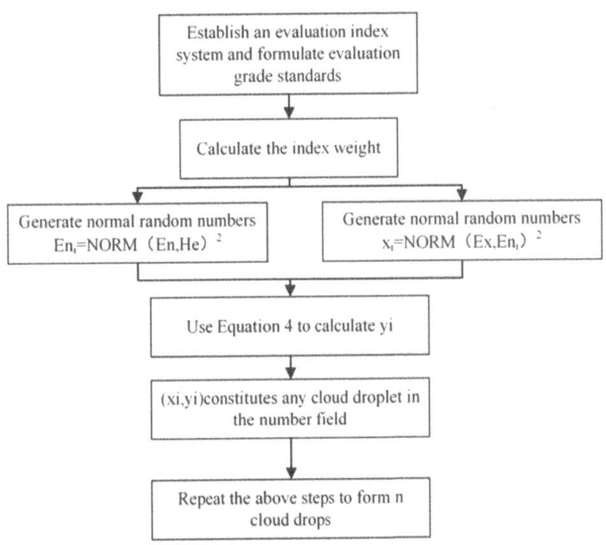

Figure 2 | Cloud model calculation flow chart.

normal random number En ', and then set Ex as the expected value and En' as the standard deviation to generate a normal random number a. After the degree of certainty is calculated by using the (Equation (4)) N cloud droplets are generated by repeating N times. The calculation principle is shown in Figure 3.

$$\frac{y = \exp\{-(x - Ex)^2}{2En'^2\}} \tag{4}$$

Determination of indicator weights

In the process of regional eco-environmental assessment, the degree of influence of different evaluation factors on the evaluation results is different. Therefore, it is first necessary to determine the degree of influence (weight) of each evaluation factor on the evaluation results. Hierarchical analysis method (AHP) is used to determine the weight of each evaluation index in the ecological environment evaluation index system. According to the steps of the analytic hierarchy process, first construct a hierarchical hierarchy structure, namely the target layer, the criterion layer and the factor layer of the evaluation index system, and then compare the two elements in turn to construct a judgment matrix, and then calculate the relative weights, and then make it consistent. The sex test is terminated after passing the test. (Changyu *et al.* 2005).

After the largest eigenvalue λ and the eigenvector $\bar{w}$ are calculated by the sum-product method, the weights of each index are acquired. And then a consistency check for each index is conducted. If the judgment matrix meets check, the calculated weight is available for use. The calculation steps are as follows:

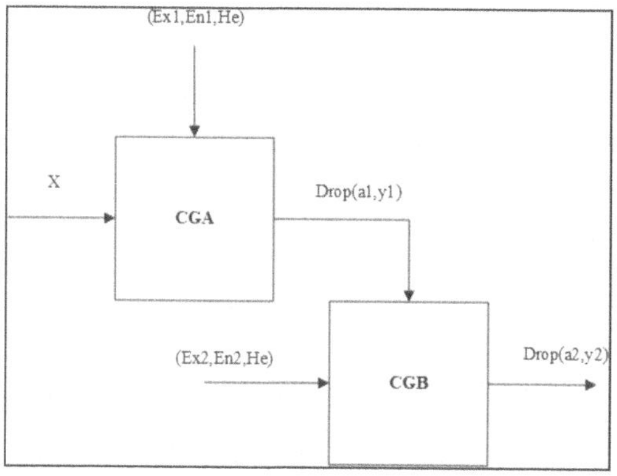

Figure 3 | Cloud generator.

- Normalize the vector

$$\bar{w}_{ij} = \frac{a_{ij}}{\sum_{i=1}^{n} a_{ij}}, j = 1, 2, 3, \ldots n \tag{5}$$

- Sum $\bar{w}_{ij}$

$$\bar{w}_{ij} = \sum_{j=1}^{n} \bar{w}_{ij}, i = 1, 2, \ldots, n \tag{6}$$

- Normalize $\bar{w}_{ij}$

$$w_i = \frac{\bar{w}_i}{\sum_{i=1}^{n} \bar{w}_i} \tag{7}$$

- Calculate the largest eigenvalue λ

$$\lambda = \frac{1}{n}\sum_{i=1}^{n}\left[\frac{(Aw)_i}{w_i}\right] \tag{8}$$

- Perform consistency check

$$CI = \frac{(\lambda - n)}{(n - 1)} \tag{9}$$

$$CR = \frac{CI}{RI} \tag{10}$$

Evaluation process

The AHP method is a combination of qualitative and quantitative methods for seeking weights. It can test and reduce the influence of subjective factors, making the analysis and evaluation more objective and scientific. At the same time, the cloud model can reduce the impact of ambiguity and randomness in the evaluation process. Coupling the AHP method with the cloud model can take advantage of the two and apply it to the evaluation of uncertain systems. The evaluation process is as follows.

Select suitable indicators for the target layer and set evaluation level standards for each indicator. Use the AHP method to acquire the weight of each indicator. After calculating the degree of certainty with the help of the three digital features by cloud model, each object is evaluated by getting the indicator value. And a membership matrix is obtained by using the cloud generator. When calculating the membership of each level, the level corresponding to the maximum membership is selected as the comprehensive evaluation result of the object to be evaluated (Figure 4).

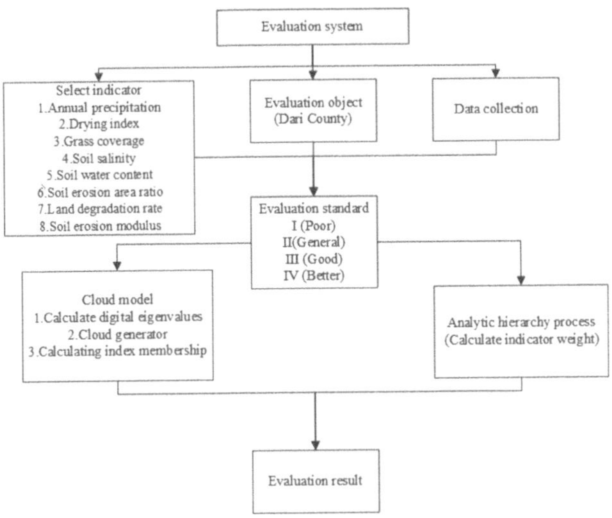

Figure 4 | Evaluation flowchart.

RESULTS AND DISCUSSION

This section introduces the calculation process of eco-environmental impact assessment based on the data of Dari County. Use the three digital feature expected values (*Ex*), entropy (*En*), and superentropy (*He*) of the normal cloud model to characterize the qualitative concepts of each index. According to Equations (1)–(3), the corresponding range of levels of each index in Table 2 is represented as a normal cloud, as shown in Table 4.

According to the evaluation standards of each index, the forward cloud generator model is used to obtain the standard normal cloud model parameters of each index by calculating *Ex*, *En*, and *He* of each index corresponding to different evaluation levels (Table 5). Then N ($N = 1{,}000$) cloud droplets are generated, and the normal membership cloud maps of each evaluation index are obtained. The abscissa represents the actual value of grassland coverage, and the ordinate represents the degree of correlation corresponding to each security level. A certain degree of association may correspond to different index values, which reflects the fuzzy uncertainty between the index value and the level to be determined (Figure 5).

The cloud certainty of each index is calculated through forward cloud generator to form a membership matrix based on the preliminary calculating results, as shown in Table 6.

The weight of each index in the ecological environment impact assessment is calculated by the above-mentioned

Table 4 | Grading standards of evaluation indexes in Dari County

	Evaluation standard			
Evaluation index	**I (Poor)**	**II (General)**	**III (Good)**	**IV (Better)**
P1	(750, 83.3, 0.01)	(1, 150, 50, 0.01)	(1, 400, 33.3, 0.01)	(1, 750, 83.3, 0.01)
P2	(5, 0.33, 0.01)	(2.6, 0.46, 0.01)	(1.1, 0.03, 0.01)	(0.5, 0.16, 0.01)
S1	(7.5, 2.5, 0.01)	(17.5, 0.83, 0.01)	(22.5, 0.83, 0.01)	(27.5, 0.83, 0.01)
S2	(6, 0.33, 0.01)	(4, 0.33, 0.01)	(2.5, 0.16, 0.01)	(1, 0.33, 0.01)
S3	(2.5, 0.83, 0.01)	(8.5, 1.17, 0.01)	(13.5, 0.5, 0.01)	(17.5, 0.83, 0.01)
R1	(42.5, 2.5, 0.01)	(27.5, 2.5, 0.01)	(12.5, 2.5, 0.01)	(2.5, 0.83, 0.01)
R2	(12.5, 0.83, 0.01)	(8, 0.67, 0.01)	(5, 0.33, 0.01)	(2, 0.67, 0.01)
R3	(5, 500, 166.67, 0.01)	(3, 750, 416.67, 0.01)	(1, 750, 250, 0.01)	(500, 166.67, 0.01)

Table 5 | Scale and meaning

Scale level	Relative importance judgment
1	X1 is just as important as X2
3	X1 is more important than X2
5	X1 is significantly more important than X2
7	X1 is much more important than X2
9	X1 is most important than X2
2,4,6,8	Between adjacent odd numbers

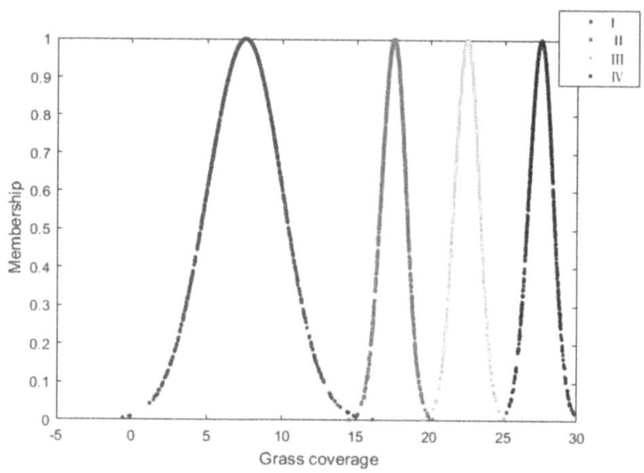

Figure 5 | Cloud model of grassland coverage in Dari County.

Table 6 | Membership of the evaluation index cloud model in Dari County

	Membership			
Evaluation index	I (Poor)	II (General)	III (Good)	IV (Better)
P1	0.09	0	0	0
P2	0	0.36	0	0
S1	0	0	0	0.12
S2	0	0.16	0	0
S3	0	0.69	0	0
R1	0	0.84	0	0
R2	0.43	0	0	0
R3	0	0.54	0	0

AHP method. The result is W = [0.1178, 0.0393, 0.0797, 0.1392, 0.0304, 0.0696, 0.3647, 0.1591].

The relationship of comprehensive membership (B), which reflects the ecological environment assessment level

Table 7 | Membership degree of evaluation index in normal cloud model in Dari County

	Evaluation standard				
Area	I (Poor)	II (General)	III (Good)	IV (Better)	Grade
Dari County	0.17	0.2	0	0	II级

of Dari County, index weight (W) and index membership (R) is shown as Equation (11).

$$B = W \times R \tag{11}$$

The four levels' comprehensive membership of ecological environment assessment in Dari County is calculated based on the formula. According to the maximum value determination method, the evaluation level belongs to 'general' (Table 7). Although the ecological environment of Dari County has been in a 'general' state, land desertification, which is one of the most serious ecological and environmental problems, cannot be underestimated. The assessment result of Dari County carried by cloud model is basically consistent with actual conditions, indicating that the assessment model has certain rationality and reliability.

CONCLUSIONS

This paper evaluates the ecological and environmental impacts in Dari County. The evaluation results intuitively reflect the ecological environment. The following conclusions are mainly drawn.

- As can be seen in the evaluation results, the ecological environment in Dari County is in a 'general' state. It urgently needs to be improved to reach a sustainable development state.
- As can be seen in the PSR index system, the land degradation rate in Dari County is particularly serious. In future work, the prevention of 'black soil beach' and the improvement of vegetation coverage, such as forest rate, in Dari County should be paid more attention, the problems of soil erosion, land salinization should be concerned, and the pressure on the local ecosystem due to human activities should be effectively controlled.

Problems in ecological environment affect many aspects such as water resources and human activities, and the characteristics of the ecological environment are quite different in different regions. Therefore, no unified understanding of ecological environment impact assessment is accepted by scholars at home and abroad. This paper analyzes the ecological environment characteristics and actual conditions of Dari County, which is in the source region of the Yellow River, constructs an evaluation system based on the PSR concept framework and builds an ecological environment impact assessment model based on the cloud model. The results provide a preparation for further study on the ecological environment of the source Yellow River region. However, there are also some deficiencies in this paper. The ecological and hydrological impact on the local ecological environment is mainly considered in this paper, and the impact of human activities, such as rodent harm, overgrazing, and other factors have not been discussed in the evaluation system.

ACKNOWLEDGEMENTS

This work is funded by the National Key R&D Program of China (2017YFC0403600, 2017YFC0403606), the National Natural Science Foundation of China (No.51779013,51639005), National Public Research Institutes for Basic R&D Operating Expenses Special Project (No.CKSF2019478, No.CKSF2017061). Special thanks are given to the anonymous reviewers and editors for their constructive comments.

DATA AVAILABILITY STATEMENT

All relevant data are included in the paper or its Supplementary Information.

REFERENCES

Bartsev, S. I., Degermendzhi, A. G., Okhonin, V. A. & Saltykov, M. Y. 2012 An integrated approach to the assessment of an ecological impact of industrial products and processes. *Procedia Environmental Sciences* **13**, 837–846.

Carlos, C., Rubén, V., Breixo, M.-R. & Alejo, C. 2020 The cotton-strip assay as an environmental surveillance tool for ecological integrity assessment of rivers affected by WWTP effluents. *Water Research* **169**, 115247.

Changyu, L., Deyi, L., Yan, D. & Xu, H. 2005 Statistical analysis of normal cloud model. *Information and Control* **02**, 236–239. +248.

He, W., Zhanxiu, Z. & Xiaoyong, W. 2018 Analysis on the trend and cycle of temperature change in Dari county from 1956 to 2015. *Modern Agricultural Science and Technology* **18**, 214–215.

Hjorth, P. & Bagheri, A. 2005 Navigating towards sustainable development: a system dynamics approach. *Futures* **38** (1), 74–92.

Jian, S., Yuhua, L., Jianglan, S., Gaihe, Y. & Dexiang, W. 2006 Comprehensive evaluation of the ecological environment quality in the source area of the river. *Journal of Northwest A & F University (Natural Science Edition)* **6** (10), 61–66.

Jing, W., Yamin, G., Lin, S., Tao, J., Xinpeng, T. & Guangde, S. 2015 Evaluation of ecological environment vulnerability in three river source regions. *Chinese Journal of Ecology* **34** (07), 1968–1975.

Jinrui, Y., Yang, F. & Fengxia, L. 2003 Research on the ecological environment division and quality evaluation index system of Qinghai Province. *Qinghai Meteorology* **7** (01), 34–40.

Jun, Y., Zhanqi, W., Gui, J., Bingyin, C. & Xianhui, H. 2013 Evaluation of post-implementation benefit of land improvement projects based on AHP and fuzzy comprehensive evaluation. *Resources and Environment in the Yangtze River Basin* **22** (08), 1036–1042.

Li, D. 1998 Uncertainty reasoning based on cloud models in controllers. *Computers and Mathematics with Applications* **35** (3), 99–123.

Long, Y., Kelong, C., Shuangcheng, L., Shengkui, C. & Yanpeng, W. 2012 Relationships between soil physical and chemical properties, species diversity, and aboveground biomass of alpine meadows in Dari county. *Territory & Natural Resources Study* **03**, 92–94.

Lujun, L. & Yaping, Y. 2000 Research on the evaluation methods of county ecological environment quality. *Environmental Monitoring Management and Technology* **04**, 13–17.

Mingmei, G., Tao, S., Tianyan, Z., Hongjun, D. & Kun, Z. 2015 Application of normal cloud model in evaluation of land ecological security in Wanjiang area. *Journal of Hunan Agricultural University (Natural Science Edition)* **41** (02), 196–201.

Qing, W. & Yanli, Z. 2002 Analysis of ecological environment changes and water resources vulnerability in the source regions of the Yellow River. *Water Resources Protection* **04**, 21–24.

Sanwei, H., Peng, P., Haijun, W. & Lian, Y. 2011 Eco-environment evaluation of agricultural land based on PSR and cloud theory: a case study of Xinxing County,

Guangdong Province. *Journal of Natural Resources* **26** (08), 1346–1352.

Yang, P., Mao, X., Li, T. & Gao, X. 2011 Ecological risk assessment of the Shenzhen River-Bay Watershed. *Human and Ecological Risk Assessment* **17** (3), 580–597.

Yang, Z., Jinming, Y., Ping, J. & Nai, Y. 2013 Evaluation of ecological security of land resources in Hubei province based on normal cloud model. *Transactions of the Chinese Society of Agricultural Engineering* **29** (22), 252–258.

Ying, X., Zeng, G.-M., Chen, G.-Q., Tang, L., Wang, K.-L. & Huang, D.-Y. 2007 Combining AHP with GIS in synthetic evaluation of eco-environment quality – a case study of Hunan Province, China. *Ecological Modelling* **209** (2), 97–109.

Zezhong, Z., Qingqing, Q., Yun, G., Gangfu, S., Xiaonan, C. & Xin, W. 2014 Comprehensive evaluation of river ecological impact based on cloud theory. *Journal of North China University of Water Resources and Hydropower (Natural Science* **35** (02), 30–34.

Zhen, L., Cao, S., Wei, Y., Dilly, O., Liu, X., Li, F., Koenig, H., Tscherning, K. & Helming, K. 2009 Comparison of sustainability issues in two sensitive areas of China. *Environmental Science and Policy* **12** (8), 1153–1167.

Zhongzheng, L., Shaopeng, Q. & Huijun, G. 2003 Comprehensive evaluation of ecological environment quality in the upper reaches of the Yellow River and its source area. *Geo-Information Science* **5** (01), 11–15.

First received 1 June 2020; accepted in revised form 15 October 2020. Available online 28 October 2020

Simulation effect evaluation of single-outlet and multi-outlet calibration of Soil and Water Assessment Tool model driven by Climate Forecast System Reanalysis data and ground-based meteorological station data – a case study in a Yellow River source

Kai Li, Yongqiang Wang, Xiaodong Li, Zhe Yuan and Jijun Xu

ABSTRACT

It is the research hotspot in the field of hydrology to apply the climate model and its downscaling data into hydrological simulations, and it is very important to evaluate the accuracy of its data. In this study, the accuracy of Climate Forecast System Reanalysis (CFSR) was evaluated from two perspectives: statistical evaluation and hydrological evaluation. In the hydrological evaluation, the applicability of CFSR in the Soil and Water Assessment Tool (SWAT) model of the Yellow River source area was studied. The results show that CFSR temperature data at the source of the Yellow River is consistent with the measured temperature data, and CFSR precipitation data overestimates precipitation. In the Yellow River source runoff simulation, the SWAT model driven by CFSR can obtain satisfactory simulation results. It does not reduce the simulation accuracy at the total outlet of the basin under the multi-outlet calibration method. It also considers the spatial differences of hydrological characteristics of each sub-basin and improve the simulation accuracy of the sub-basin simulation.

Key words | CFSR, multi-outlet calibration, single-outlet calibration, SWAT model, Yellow River source

Kai Li
Yongqiang Wang (corresponding author)
Xiaodong Li
Zhe Yuan
Jijun Xu
Water Resources Department (Transboundary River Management Department),
Changjiang River Scientific Research Institute,
Wuhan 430010,
China
and
Hubei Provincial Key Laboratory of Basin Water Resources and Ecological Environmental Sciences,
Changjiang River Scientific Research Institute,
Wuhan 430010,
China
E-mail: *wangyq@mail.crsri.cn*

Xiaodong Li
College of Hydrology and Water Resources,
Hohai University,
Nanjing 210098,
China

HIGHLIGHTS

- Simulation effect of SWAT model was evaluated.
- Accuracy of precipitation and temperature data of CFSR data set was evaluated.
- Simulation effects of SWAT model and multi-outlet calibration method were compared.
- The CFSR data did not achieve good simulation effect.
- The spatial difference of the sub-basin was taken into account in the multi-outlet calibration method.

INTRODUCTION

The global climate system has been undergoing significant changes since the 1900s, and the increasingly intense human activities affect the process of hydrological cycles to varying degrees (Yang *et al.* 2020). Therefore, the hydrological cycle and the vulnerability of water resources against the backdrop of climate changes have become a focus of hydrological research. Hydrological modeling, which is based on mathematical principles and hydrometeorological data to simulate the complex hydrological cycles in nature, is a necessary and important method for hydrological research (Ouermi

doi: 10.2166/ws.2020.312

et al. 2019). Building a hydrological model driven by ground-based meteorological station (GMS) can provide a strong support for simulation of regional water cycles. High-precision and high-quality meteorological data can reduce the uncertainty of the model, thereby increasing the accuracy of simulation and prediction results (Barnett *et al.* 2008).

Conventional methods of hydrological research use the meteorological data collected from GMS to drive the hydrological model to simulate hydrological cycles and perform flood forecasting (Arnold *et al.* 1998; Golmohammadi *et al.*2014). However, limited by financial investment and monitoring technology, GMS is rare and unevenly distributed. In remote areas and inaccessible rivers, the lack of GMS leads to challenges in collecting effective meteorological data and subsequent hydrological modelling. To solve this problem, scholars all over the world have used the output data of the climate models and their downscaling data in large-scale hydrological simulations and achieved good results (Smith & Kummerow 2013). However, due to the low resolution of climate models and reanalysis data, large deviations occur when these data are applied to simulation of regional climate change with complex underlying surface features (Liu *et al.* 2017). Therefore, in hydrological response research, it is important to study the application effect and sensitivity of different meteorological reanalysis data and climate model products in the Soil and Water Assessment Tool (SWAT) model. At present, the China Meteorological Assimilation Driving Datasets (CMADS) for the SWAT model established by Meng Xianyong is widely used in China (Meng & Wang 2017). The application of this data set in several basins in China shows that its simulation results are better than those achieved by the conventional method, which is based on data from GMS. Nowadays, the CMADS has become a mature data set for hydrological studies. On the other hand, the Climate Forecast System Reanalysis (CFSR) data set, developed by the National Centers for Environmental Information (NCEI) of the U.S., has shown in recent years that its simulation effect is as good or better than that of data from GMS. However, some scholars point out that this data set fails to deliver good simulation results in certain regions, especially the tropical basins. They suggest that CFSR data be used in regions where conventional GMS are available.

The SWAT model is a semi-distributed hydrological model developed by the U.S. Department of Agriculture. It is mainly used to simulate and evaluate the hydrological situation and water quality changes of basins under various management measures and climate changes. To judge whether the SWAT model is applicable to a basin, it is necessary to calibrate the parameters of the model first. At present, the SWAT model parameter calibration methods mainly include single-station calibration and multi-station calibration. The method of single-station calibration is to set a total basin outlet, and the parameters are assumed the same throughout the whole basin (Thavhana *et al.* 2018). This method ignores the uniqueness of sub-basins, and thus the calibrated parameters cannot describe the characteristics of the corresponding sub-basin. On the other hand, the multi-station calibration method can show the spatial differences among sub-basins and improve the simulation accuracy of each sub-basin without impairing the simulation accuracy at the whole basin (Yu *et al.* 2014).

To this end, this paper takes the source of the Yellow River as the research area and evaluates the CFSR data set from two aspects: statistical evaluation and hydrological evaluation. In hydrological evaluation, CFSR data is used to drive the SWAT model. To reasonably evaluate product characteristics and reduce the impact of hydrological model calibration methods on product performance, runoff is simulated under single-outlet and multiple-outlet scenarios, and the simulation accuracy is evaluated.

RESEARCH AREA AND MATERIALS

Overview of the research area

Originating from Zhaqu of Chahasila Mountain in the Bayan Kala Mountains of Qinghai–Tibet Plateau, the Yellow River from eastwards through nne provinces in China before entering the Bohai Sea. The catchment area above Tangnaihai Hydrological Station is the source of the Yellow River (Figure 1). The catchment area covers $12.2 \times 10^4\,km^2$, accounting for 16.2% of the total area of the Yellow River Basin. The geographical range is 95° 30′–103° 30′E and 32°10 ′–36°05N. The terrain is high in the west and low in the east, with the lowest altitude of 2,546 m and the highest altitude of

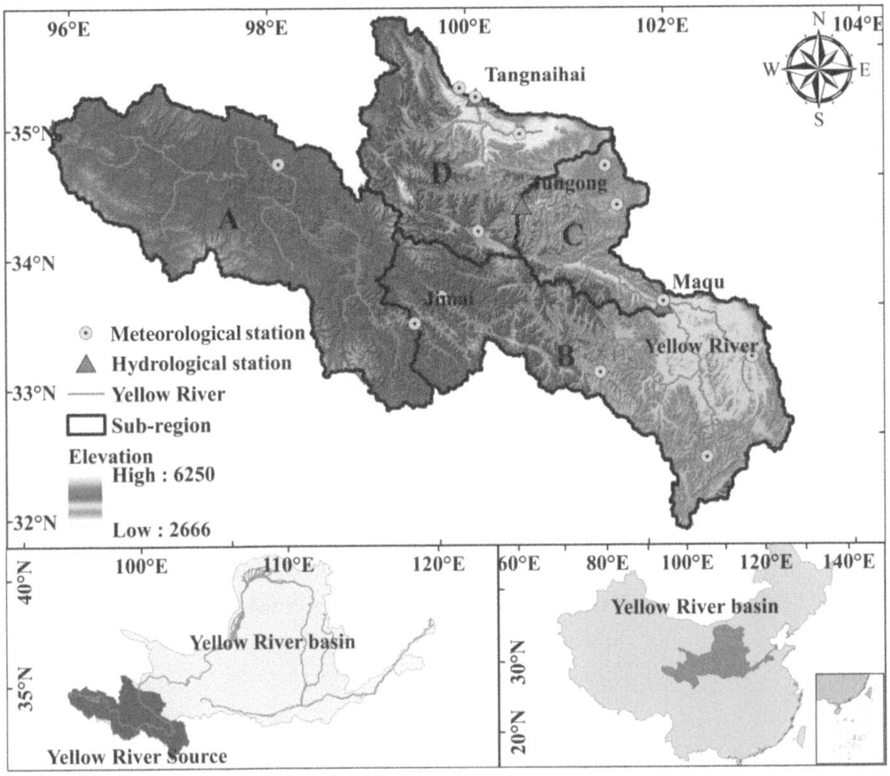

Figure 1 | Meteorological stations, hydrological stations and their control sub-regions in the source area of the Yellow River.

6,282 m (the Aemye Ma-chhen Range). The source area belongs to the sub-cold and semi-arid area in the plateau, with a high temperature in the southeast and a low temperature in the northwest, and the annual precipitation of between 287.49 mm and 754.36 mm decreases from southeast to northwest. There are many tributaries, well-developed glacial landforms, and more than 40 glaciers of various sizes. The source area is 1,959 km long, and the natural runoff is 20.52 billion m^3.

Spatial data

The spatial data, digital elevation digital elevation model (DEM) data, land use and soil data, and DEM data required for the present study were obtained from ASTER GDEM with a resolution of 30 m (Li & Zhao 2018). The land use data was from 2010 and 2015 and was obtained from the Resource and Environmental Science Data Center of the Chinese Academy of Sciences, with an accuracy is 30 m. Soil data was from the Chinese soil data set based on the Harmonized World Soil Database (HWSD).

Hydrometeorological data

In hydrological models, meteorological data input is the key factor affecting the result of runoff simulation. GMS required for the SWAT model includes daily precipitation, temperature, relative humidity, wind speed, and solar radiation (Zhang *et al.* 2005). In this research, SWAT models were driven by two data sets: GMS data set and CFSR data set. The GMS data set is derived from the China Meteorological Data Network. GMS data from nine GMS, including Xinghai, Dari, and Henan, which are located in the source area of the Yellow River, were selected as input into the SWAT model with a time span from 1997 to 2013. On the other hand, CFSR data, a third-generation analytical product developed by the NCEI of the USA, is a global, highly variable, coupled atmospheric–ocean–land–sea surface–sea ice system designed to provide best estimates of these coupling states over the period. In the present study, data from 260 CFSR stations in the source area ranging from 95°30′–103°30″ E and 32°10′–36°05′ N were used, with a spatial resolution of 38 km and a time span from 1997 to 2013.

RESEARCH METHOD

Construction and evaluation method of SWAT model

The SWAT model is a semi-distributed hydrological model developed by the U.S. Department of Agriculture in the 1990s. The model is mainly used to simulate and evaluate the hydrological process and water quality changes of basins under various management measures and climate change conditions. The simulation of the hydrological process is divided into two parts: simulation of the land surface water cycle and the confluence calculation. The water cycle process simulated by the model follows the water balance equation:

$$SW_t = SW_0 + \sum_{t=1}^{t} (R_i - Q_i - ET_i - W_i - QR_i) \tag{1}$$

where SW_t represents the final soil water content (mm), SW_0 represents the initial soil water content, t is time (day), R_i represents precipitation (mm), Q_i represents the surface runoff (mm), ET_i represents evaporation (mm), and W_i and QR_i represent water content (mm) entering the aeration zone and regression flow (mm), respectively.

Tangnaihai Hydrological Station was set as the basin outlet of the source of the Yellow River, and 54 sub-basins were divided by the SWAT model. After reclassifying the land-use types in the source area according to the SWAT land-use classification system, six types of land-use were obtained. Based on the Chinese soil data set in HWSD, 26 soil types in the source area were extracted and the required soil parameters were calculated using Soil–Plant–Air–Water (SPAW) (Jiang *et al.* 2018).

The correlation coefficient (R^2), the Nash–Sutcliffe model efficiency (NSE) coefficient and Percent Bias (PBIAS) were selected to evaluate the performance of the SWAT model in simulating runoffs. The specific calculation formula is as follows:

$$R^2 = \left[\frac{\sum_{i=1}^{n} (O_i - S_i)(S_i - \overline{S})}{(\sum_{i=1}^{n} (O_i - \overline{O})^{0.5})(\sum_{i=1}^{n} (S_i - \overline{S})^2)^{0.5}} \right] \tag{2}$$

$$\mathrm{NSE} = 1 - \frac{\sum_{i=1}^{n} (O_i - S_i)^2}{\sum_{i=1}^{n} (O_i - \overline{O})^2} \tag{3}$$

$$\mathrm{PBIAS} = \frac{\sum_{i=1}^{n} (O_i - S_i) \times 100}{\sum_{i=1}^{n} O_i} \tag{4}$$

where O_i is the measured runoff sequence, S_i is the simulated runoff sequence, $\bar{O}$ is the measured average runoff, $\bar{S}$ is the simulated average runoff, and n is the series length. It is generally considered that if $R^2 > 0.6$, NSE > 0.5, and PBIAS is <25%, the performance of the model to simulate daily runoff is satisfactory. In monthly runoff simulation, if $R^2 > 0.7$, NSE > 0.55 and PBIAS is <25%, the performance of the monthly runoff model is considered to be satisfactory (Moriasi *et al.* 2015).

Evaluation on CFSR data

In order to fully quantify the accuracy of CFSR data, continuous statistical indexes were used to assess the accuracy of monthly temperature and precipitation of CFSR data set. The correlation coefficient (CC), mean error (ME), root mean square error (RMSE), and mean bias error (MBE) were selected as evaluation indexes in the present study. CC represents the degree of linear correlation between CFSR and GMS data; ME and RMSE represent the average error of CFSR data; and MBE describes the systematic bias of data (Hu *et al.* 2014; Tan *et al.*2015). The formulas for calculating each index are as follows:

$$\mathrm{CC} = \frac{\sum_{i=1}^{n} (G_i - \bar{G})(O_i - \bar{O})}{\sum_{i=1}^{n} (G_i - \bar{G}) \sum_{i=1}^{n} (O_i - \bar{O})} \tag{5}$$

$$\mathrm{RMSE} = \sqrt{\frac{\sum_{i=1}^{n} (G_i - O_i)^2}{n}} \tag{6}$$

$$\mathrm{ME} = \frac{\sum_{i=1}^{n} (O_i - G_i)}{n} \tag{7}$$

$$\mathrm{MBE} = \frac{\sum_{i=1}^{n} (O_i - G_i)}{\sum_{i=1}^{n} O_i} \quad (8)$$

where G_i is the temperature/precipitation data in CFSR; O_i is the temperature/precipitation data in GMS; $\bar{G}$ is the average temperature/precipitation data in CFSR; $\bar{O}$ is the average temperature/precipitation data in GMS; and n is the data length.

Simulation method

Single-outlet simulation

Single-outlet simulation means that the model only has one outlet for the calibration of the model's parameters, so that the model parameters of the entire basin are consistent. Specific implementation steps, from upstream to downstream, sequentially rate the hydrological stations. When the Tangnaihai Hydrological Station is calibrated, the parameter values of all sub-basins in the source area of the Yellow River are consistent.

Multi-outlet simulation

Multiple-outlets are set in the basin and each sub-basin can be regarded as a single-outlet. The variation of parameters in each sub-basin is related to the hydrological characteristics of the sub-basin. Even if the same parameters are used, their changes in each sub-basin are inconsistent. First, as shown in Figure 1, area A is calibrated using the Jimai Hydrological Station to obtain a set of optimal parameters for the area. When calibrating the Maqu Station, the control drainage area at this time is A + B, but the parameters of area A have been calibrated at the Jimai Station. The parameters of area A are not changed, and only area B is calibrated. Then, when calibrating the Jungong Hydrological Station, its control basin area is A + B + C, the previous A and B have been calibrated, the parameters are unchanged, and only are C is calibrated. calibrating The Tangnaihai Hydrological Station, which controls the basin area A + B + C + D, is then calibrated. Similarly, A, B, and C do not need to be rated, only area D needs calibrated. Finally, the parameters of the four sub-regions A, B, C, and D are different.

RESULTS AND DISCUSSION

Parameter sensitivity analysis

The SWAT model contains many parameters that affect the results of runoff simulation. Calibrating all parameters can lead to excessive uncertainties and excessive parameters. Therefore, parameter sensitivity analysis is required to improve the efficiency and accuracy of calibration. The global sensitivity analysis of SWAT calibration and uncertainty program (SWAT-CUP) is used to analyze all parameters that affect the runoff. According to t-stat and *P*-value, the greater the absolute value of t-stat and the closer the *P*-value to 0, the more sensitive the parameters are to the model. In this study, 12 parameters with high sensitivity are selected for model calibration (Table 1).

Evaluation on data accuracy

Valuation of temperature and precipitation data

In order to assess the accuracy of CFSR temperature data in the source area of the Yellow River, GMS data were used to verify the accuracy of CFSR data. In this paper, hydrometeorological data from nine GMSs were used to drive the SWAT model. We chose four representative stations, and selected CFSR stations with similar coordinates based on their latitude and longitude for evaluation. The CC, RMSE, ME, and MBE were used as evaluation indexes. The specific evaluation results are shown in Table 2.

Table 2 reveals that the CC value of the CFSR temperature data at four stations is close to the ideal value 1, which indicates that the CFSR temperature data is closely related to the GMS temperature data. ME values at the highest temperatures at all stations were positive and MBE values were >0, indicating that CFSR data underestimated

Table 1 | Results of model sensitivity analysis

Parameter	Parameter significance	GMS t-stat	GMS *P*-value	GMS Ranking	CFSR t-stat	CFSR *P*-value	CFSR Ranking
SMTMP	Average slope length	10.85	0.000	1	5.59	0.000	3
CN2	Soil Conservation Service runoff curve number	−5.79	0.000	2	−4.35	0.000	4
SMTMP	Melting temperature	−5.55	0.000	3	−10.66	0.000	1
SFTMP	Rain-snow boundary temperature	−5.25	0.000	4	−6.37	0.000	2
ALPHA_BF	Baseflow recession constants	−4.10	0.000	5	−3.70	0.000	6
GWQMN	Return threshold of shallow groundwater	4.05	0.000	6	4.05	0.000	5
SOL_K	Soil saturated hydraulic conductivity	−2.90	0.003	7	−2.90	0.003	7
CH_K2	Effective hydraulic conductivity in main river channel	2.79	0.005	8	2.79	0.005	8
GW_REVAP	Reevaporation coefficient of shallow groundwater	2.53	0.011	9	2.53	0.011	9
CH_N2	N value of Manning's formula in river channel	2.13	0.033	10	2.13	0.033	10
EPCO	Compensation coefficient of plant transpiration	2.02	0.043	11	−2.02	0.043	11
SURLAG	Hysteresis coefficient of surface runoff	1.76	0.078	12	1.76	0.078	12

Table 2 | CFSR data evaluation

		Maqin (3451003)	Dari (339997)	Maqu (3391022)	Hongyuan (3291025)
Highest temperature	CC	1	1	0.99	0.98
	RMSE	6.9	5.5	3.3	4.1
	ME	5.9	4.5	1.7	2.8
	MBE	0.63	0.57	0.17	0.25
Lowest temperature	CC	1	0.99	0.99	0.98
	RMSE	4.12	3.45	3.2	5.05
	ME	−1.2	0.44	−1.5	−3.29
	MBE	0.18	−0.07	0.49	0.72
Precipitation	CC	0.59	0.45	0.44	0.49
	RMSE	5.1	3.71	5.2	4.7
	ME	−1.9	0.038	−1	−0.75
	MBE	−1.35	0.024	−0.67	−0.38

Note: The corresponding CFSR station number in brackets represents the meteorological station.

the temperature. The average ME of the lowest temperatures at four stations were above 0, and the MBE mean >0, indicating that CFSR data overestimated the lowest temperatures.

As for precipitation data, the CC values of the daily CFSR data at all four stations were more than 0.4, which indicates that a weak correlation between CFSR precipitation data and GMS precipitation data. The mean values of ME and MBE of CFSR data from four stations are <0, which indicates that CFSR data overestimated four meteorological precipitation data in the source area of the Yellow River.

Evaluation of accuracy of the SWAT model

Single-outlet simulation

According to the actual measured hydrological data sequence and in order to make all the hydrological processes at the initial stage of the simulation go from the initial state to the equilibrium state, a 3-year warm-up period was set. The years 2000–2009 was set as the model calibration period and 2010–2013 as the validation period. Using the combination of the SUFI-2 optimization algorithm in SWAT-CUP2012 and manual calibration, the 12 main parameters in the model were calibrated. Figure 2 shows the runoff simulation results during the model calibration period (2000–2013).

Figure 2 shows that the simulation results of the SWAT model constructed by GMS data have a high fitting degree, while the SWAT model constructed by CFSR data has a lower fitting degree. At the same time, from the index evaluation results of the four stations in Table 3, it can be seen that values of R^2 and NSE of the SWAT model constructed by GMS data are generally above 0.75, and PBIAS is <16.1% in the calibration and validation period, and the simulation results of this model are satisfactory. On the other hand, it can be seen that the R^2 of the model built by CFSR in the calibration period is above 0.68, the NSE value is above 0.65, and the PBIAS coefficient is lower than 10.4%. Except for the Jimai Station, the R^2 and NSE of the other stations during the validation period are both greater than 0.75, which indicates that the simulation results of the SWAT model driven by GMS are basically satisfactory. However, regardless of the kind of data the model is based on, the simulation results in the downstream are obviously better than those in the upstream, mainly because the regulation and storage functions of lakes were not considered in this study. There are two large lakes in the upper reaches of the source area of the Yellow River–Zhaling Lake, and Eling Lake, which have a regulating and storing effect on runoff. According to existing studies, the retention and regulation of the two lakes make the annual runoff distribution of the Yellow River, along with the source of the Yellow River, more uniform (Li *et al.* 2001). However, the study found that the runoff from the Huangheyan Hydrological Station only accounted for 19%, 6%, 5%, and 4% from Jimai, Maqu, Jungong, and Tangnaihai, respectively. Therefore, the influence from Zhaling Lake and Eling Lake on the runoff to the lower Yellow River gradually decreases. In general, the SWAT model results constructed with GMS data are better than those using CFSR data in the source area of the Yellow River.

Multi-outlet simulation

In this paper, considering the geographical location of hydrological stations and effective hydrological data, the source area of the Yellow River is generalized into four sub-regions (Figure 1), and the hydrological stations at the outlet of each region are set as the calibration points of the region. From upstream to the outlet of the basin, the hydrological stations are successively Jimai (A), Maqu (B), Jungong (C), and Tangnaihai (D). The four stations constitute nine multi-outlet calibration situations (Table 4), and

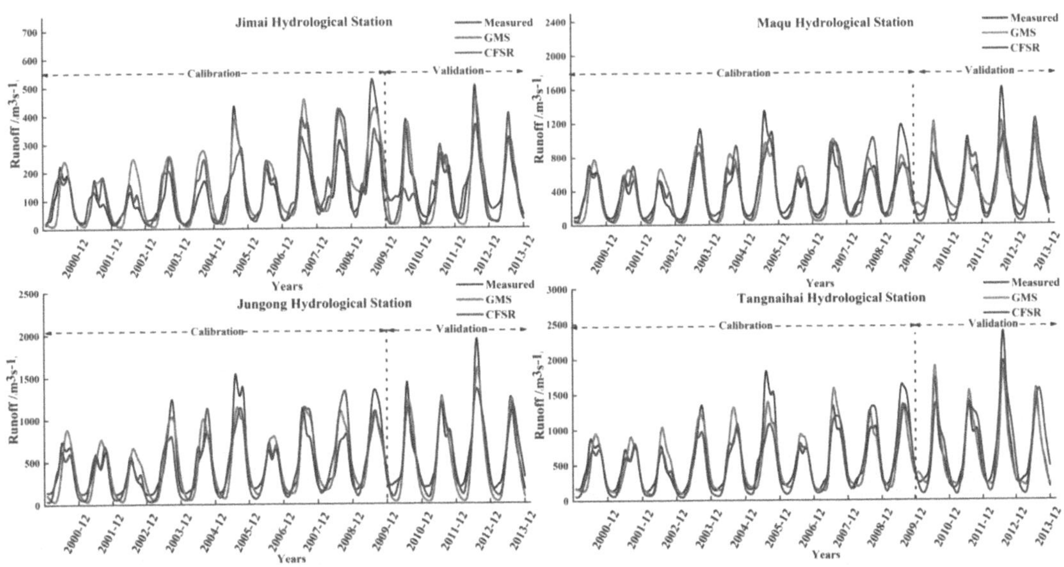

Figure 2 | GMS and CFSR calibration results in the single-outlet.

Table 3 | Evaluation results of single-outlet simulation effect

Period	Station	GMS			CFSR		
		R^2	NSE	PBIAS %	R^2	NSE	PBIAS %
Calibration period (2000–2009)	Jimai	0.75	0.65	−20.1	0.68	0.65	−7.2
	Maqu	0.76	0.72	10.9	0.72	0.71	−0.9
	Jungong	0.74	0.73	−0.2	0.73	0.71	2.7
	Tangnaihai	0.77	0.76	3.2	0.79	0.77	10.4
Validation period (2010–2013)	Jimai	0.79	0.68	15.4	0.59	0.60	16.9
	Maqu	0.88	0.86	−1.8	0.75	0.75	1.5
	Jungong	0.86	0.80	16.1	0.81	0.80	4.7
	Tangnaihai	0.90	0.89	6.2	0.77	0.75	−6.4

Table 4 | Multi-outlet calibration situation

Situation	Calibration station	Comparable station
2 control stations	Jimai/Maqu	Maqu
3 control stations	Jimai/Maqu/Jungong	Jungong
4 control stations	Jimai/Maqu/Jungong/Tangnaihai	Tangnaihai

the calibration results are compared with the single-outlet calibration.

According to Figure 3, the simulation results are slightly worse than the single-outlet simulation results. The difference is mainly manifested in the overestimation of peak flow in 2000 and 2003 and the flow process is advanced. However, during the verification period, the multi-outlet calibration method is generally better than the single-outlet calibration method, and the simulation results fit the measured runoff process line to a higher degree. From the evaluation indicators of the model in Table 5, it can be seen that during the calibration period, both R^2 and NSE indicators are significantly reduced, especially at the Maqu Station, R^2 is reduced by 18%, and NSE is reduced by

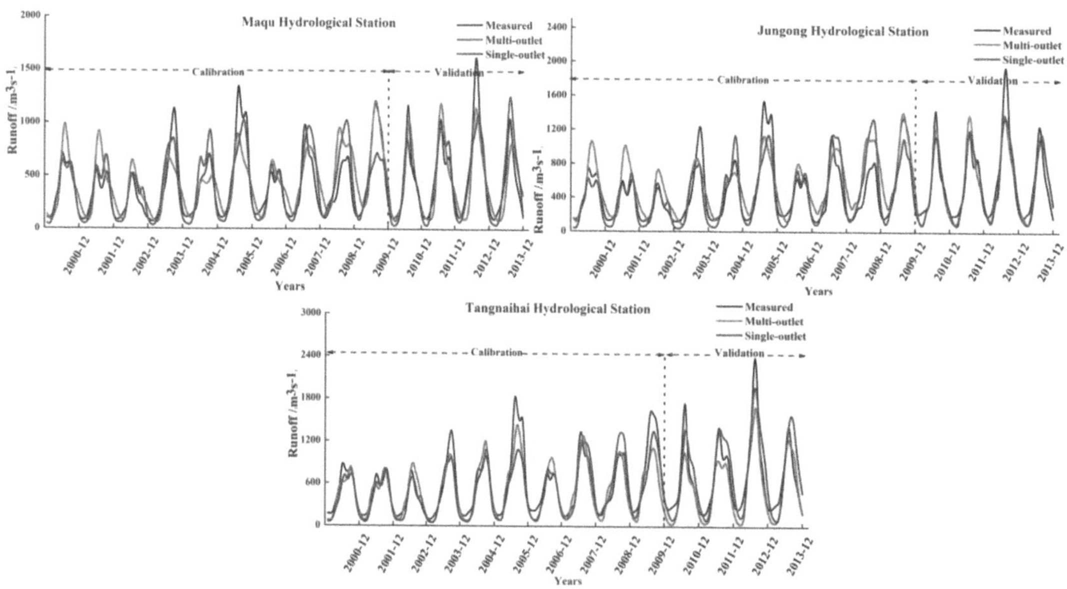

Figure 3 | Comparison of single-outlet and multi-outlet of three hydrological stations.

Table 5 | Evaluation results of single-outlet and multi-outlet simulation effect

Period	Station	Single-outlet			Multi-outlet		
		R^2	NSE	PBIAS %	R^2	NSE	PBIAS %
Calibration period (2000–2009)	Jimai/Maqu	0.72	0.71	−0.9	0.59	0.56	−8.1
	Jimai/Maqu/Jungong	0.73	0.71	2.7	0.66	0.61	−16.9
	Jimai/maqu/Jungong/Tangnaihai	0.79	0.77	10.4	0.78	0.76	9
Validation period (2010–2013)	Jimai/Maqu	0.75	0.75	1.5	0.77	0.77	5.7
	Jimai/Maqu/Jungong	0.81	0.8	4.7	0.8	0.79	6.2
	Jimai/Maqu/Jungong/Tangnaihai	0.77	0.75	−6.4	0.83	0.74	23.5

21%. During the verification period, the R^2 and NSE indicators were slightly improved through the multi-outlet calibration method: under the multi-outlet calibration at Maqu Station, R^2 and NSE increased by 1% and the R^2 at Tangnaihai Station increased by 5%. In general, the multi-outlet simulation does not reduce the simulation accuracy of the basin total outlet, and it also enables the spatial differences of each sub-basin hydrological characteristics to improve the simulation accuracy of the sub-basin simulation.

As shown in Table 6, after multi-outlet calibration, the model parameters have changed significantly. Most of the parameter values are different to those under single-outlet calibration. For example, when the parameter GWQMN is in single-outlet calibration mode, the value of this parameter is proportional to the simulation effect, the larger the value, the better the simulation effect. In the multi-outlet calibration mode, the parameter value is inversely proportional to the stimulation effect, and the change of each parameter is more in line with the underlying surface characteristics of the sub-basin.

CONCLUSION

This article evaluates the applicability of the CFSR data set in the source region of the Yellow River from both statistical and hydrological aspects. In the hydrological evaluation, in order to reasonably evaluate the data and reduce the impact of the SWAT model calibration method on the simulation results, single-outlet simulations and multi-outlet simulations were carried out.

Table 6 | The parameter fitted value of single-outlet and multi-outlet simulation effect

Parameter name	Maqu		Jungong		Tangnaihai	
	Single-outlet	Multi-outlet	Single-outlet	Multi-outlet	Single-outlet	Multi-outlet
CN2	59.68	51.78	25.78	51.75	48.18	61.27
ALPHA_BF	0.27	0.07	0.47	1.00	0.43	0.07
GWQMN	3,703.59	4,463.84	3,855.93	3,328.98	5,000.00	0.00
GW_REVAP	0.20	0.20	0.13	0.12	0.12	0.12
EPCO	0.20	0.10	0.11	0.97	0.96	1.00
SLSUBBSN	87.63	128.38	91.58	75.22	38.92	113.65
SFTMP	−20.00	−1.34	−15.69	−9.05	−25.76	4.20
SURLAG	0.00	23.96	6.99	24.00	23.00	5.69
SOL_ALB	0.04	0.06	0.07	0.23	−0.03	0.13
CH_N2	0.30	0.19	0.30	0.23	0.19	−0.01
CH_K2	405.32	268.96	402.13	311.82	500.00	30.91

According to the data statistical evaluation, it is found that the CFSR temperature data at the source of the Yellow River is basically consistent with the measured temperature data, but the temperature is overestimated. CFSR precipitation data has a general linear relationship with GMS precipitation data, but the precipitation is also overestimated.

This paper used CFSR data and GMS data to drive the SWAT model, and compared model runoff simulation results at Jimai, Maqu, Jungong, and Tangnaihai Hydrological Stations. It found that both data set-driven models can achieve satisfactory simulation results. In general, the simulation effect of GMS is better than that of CFSR data.

The simulation result of the CFSR data set under the multi-outlet calibration was not as good as the single-outlet calibration method. However, during the verification period, the simulation result of the multi-outlet calibration method was worse than the single-outlet calibration method. In addition, the multi-outlet calibration method did not reduce the simulation accuracy of the total outlet of the basin, but it can consider the spatial differences of the hydrological characteristics of each sub-basin to improve the simulation accuracy of the sub-basin simulation.

ACKNOWLEDGEMENT

This research is supported by the National Key R&D Program of China (2017YFC0403600, 2017YFC0403606), the National Natural Science Foundation of China (No. 51779013, 51639005), National Public Research Institutes for Basic R&D Operating Expenses Special Project (No. CKSF2019478, No.CKSF2017061) and the Postgraduate Research & Practice Innovation Program of Jiangsu Province (SJCX20_0158). Special thanks are given to the anonymous reviewers and editors for their constructive comments.

DATA AVAILABILITY STATEMENT

All relevant data are included in the paper or its Supplementary Information.

REFERENCES

Arnold, J. G., Srinivasan, R., Muttiah, R. S. & Williams, J. R. 1998 Large area hydrologic modeling and assessment part I: model development. *Journal of the American Water Resources Association* **34** (1), 73–89.

Barnett, T. P., Pierce, D. W., Hidalgo, H. G., Bonfils, D., Das, T., Bala, G., Wood, A. W., Nozawa, T., Mirin, A. A., Cayan, D. R. & Dettinger, M. D. 2008 Human-induced changes in the hydrology of the western United States. *Science* **319** (5866), 1080–1083.

Golmohammadi, G., Prasher, S., Madani, A. & Rudra, R. 2014 Evaluating three hydrological distributed watershed models: MIKE-SHE, APEX, SWAT. *Hydrology* **1** (1), 20–39.

Hu, Q., Yang, D., Li, Z., Mishra, A. K., Wang, Y. & Yang, H. 2014 Multi-scale evaluation of six high-resolution satellite monthly rainfall estimates over a humid region in China with dense rain gauges. *International Journal of Remote Sensing* **35** (4), 1272–1294.

Jiang, X. F., Wang, L., Ma, F., Li, H. Q., Zhang, S. J. & Liang, X. W. 2018 Localization method for SWAT model soil database based on HWSD. *China Water & Wastewater* **30** (11), 135–138.

Li, H. & Zhao, J. 2018 Evaluation of the newly released worldwide AW3D30 DEM over typical landforms of China using two global DEMs and ICESat/GLAS data. *IEEE Journal of Selected Topics in Applied Earth Observations and Remote Sensing* **11** (11), 4430–4440.

Li, W. S., Feng, L. & Shun, S. L. 2001 Influence of Zhaling and Eling lake on the annual discharge of the Huanghe River source area. *Acta Geographica Sinica* **56** (1), 75–82.

Liu, J., Liu, S. Y., ShangGuan, D. H. & Xu, J. D. 2017 Applicability evaluation of precipitation datasets from CMADS, ITPCAS and TRMM 3B42 in Yurungkax River Basin. *Journal of North China University of Water Resources and Electric Power (Natural Science Edition)* **38** (05), 28–37.

Meng, X. Y. & Wang, H. 2017 Significance of the China meteorological assimilation driving datasets for the SWAT model (CMADS) of East Asia. *Water* **9** (10), 765.

Moriasi, D., Gitau, M. N., Pai, N. & Daggupati, P. 2015 Hydrologic and water quality models: performance measures and evaluation criteria. *Transactions of the American Society of Agricultural and Biological Engineers* **58** (6), 1763–1785.

Ouermi, K. S., Paturel, J.-E., Adounpke, J., Lawin, A. E., Goula, B. T. A. & Amoussou, E. 2019 Comparison of hydrological models for use in climate change studies: a test on 241 catchments in West and Central Africa. *Comptes Rendus – Geoscience* **351** (7), 477–486.

Smith, R. A. & Kummerow, C. D. 2013 A comparison of in situ, reanalysis, and satellite water budgets over the Upper Colorado River Basin. *Journal of Hydrometeorology* **14** (3), 888–905.

Tan, M., Ibrahim, A., Duan, Z., Cracknell, A. & Chaplot, V. 2015 Evaluation of six high-resolution satellite and ground-based precipitation products over Malaysia. *Remote Sensing* **7** (2), 1504–1528.

Thavhana, M. P., Savage, M. J. & Moeletsi, M. E. 2018 SWAT model uncertainty analysis, calibration and validation for runoff simulation in the Luvuvhu River catchment, *South Africa. Physics and Chemistry of the Earth, Parts A/B/C* **105**, 115–124.

Yang, Y. H., Weng, B. S., Man, Z. H., Yu, Z. L. & Zhao, J. L. 2020 Analyzing the contributions of climate change and human activities on runoff in the Northeast Tibet Plateau. *Journal of Hydrology: Regional Studies* **27**, 100639.

Yu, D., Dong, X. H., Li, L., Wang, J. c. & Li, Z. H. 2014 Comparison of multi-gage and single-gage calibration of SWAT model for runoff simulation in Qingjiang River Basin. *Water Resources and Power* **32** (05), 14–18.

Zhang, W. J., Tang, Y. & Wang, H. Q. 2005 Estimation of daily solar radiation in China. *Chinese Journal of Agrometeorology.* **26** (3), 165–169.

First received 10 August 2020; accepted in revised form 26 October 2020. Available online 11 November 2020

Research on water resources dispatch model based on improved genetic algorithm – water resources dispatch model

Haoran Fu and Huahui Li

ABSTRACT

According to the research on reservoir water resources scheduling and distribution, the aim is to balance the water supply and demand in each period, and consider the total water supply and the annual external water withdrawal of the reservoir in each period as water rights. The decision-making variables are provided for the water supply of the reservoir in this paper, so that water demand of the water-receiving area can be better met to alleviate the water shortage at various stages and realize the effective use of water resources. Moreover, through the constraints of reservoir operation rules and other constraints, a mathematical model is established for optimal operation of water resources in the reservoir system. Meanwhile, optimized genetic algorithms are applied to solve the model according to the characteristics of the model. After simulation tests, compared with the traditional linear binary algorithm used in the reservoir, the improved genetic algorithm studied in the paper improves the accuracy of data calculation and data convergence, which proves that the research results of the paper provide theoretical and practical significance for improving the level of reservoir water resources management and solving the problem of optimal water resources scheduling.

Key words | improved genetic algorithm, reservoir index, water resources dispatch, water resources model

Haoran Fu
School of Economics,
Anyang Normal University,
Anyang, Henan 455000,
China

Huahui Li (corresponding author)
Institute of Civil Engineering,
Anyang University,
Anyang, Henan 455000,
China
E-mail: *lhh1883721@163.com*

HIGHLIGHTS

- This paper realizes the supply decision variables.
- The data index takes the historical data of the past as a reference to compare the algorithm.
- The solution model of this algorithm makes use of the genetic effect of the improved genetic algorithm.
- The constraint conditions are used as the rules of data allocation.
- A multi-group genetic algorithm based on real coding is proposed.

INTRODUCTION

In the field of water environment planning and management, a large number of parameters usually have observational subjectivity and unreliability problems. In addition, according to the linear binary algorithm data used for reservoir water resources scheduling, the uncertainty of parameters is often described in a fuzzy manner. Therefore, in order to ensure the existence of the optimal solution to the planning problem, there is usually a factor

doi: 10.2166/ws.2020.344

gap between one index parameter and another index, and the upper and lower boundaries of the gap will be given. In practical problems, due to the many parameters and uncertainties involved in the aquatic ecological environment (Chai *et al.* 2017a, 2017b), the genetic system theory is usually used to construct a water environment management model to solve non-deterministic planning problems.

In the paper, genetic algorithms are adopted to treat each data index as a genetic gene, and each chromosome represents an independent solution in the entire group. Additionally, the chromosome is composed of a binary combination string (Ai *et al.* 2017), which represents the elimination efficiency of the reservoir and the water supply of different water sources. Although a linear binary algorithm can be used to solve the unconstrained production problem, the method is based on a penalty function rather than constraints, and the penalty solution exceeds the constraints given by the system. Therefore, a fitting function is established to improve the genetic genes in the paper (Chai *et al.* 2017a, 2017b). The paper outlines modern optimization methods for evolutionary search, focusing on applications in the areas of water resources planning, engineering, and management (Tayfur 2017). In the view of traditional genetic algorithm problem, a fitting function is established to treat each data index as a genetic gene, and a multi-group genetic algorithm based on real coding is proposed to simplify the genetic calculation process while effectively avoiding premature immature convergence problems, which provides conditions for the solution.

OPTIMIZED DISPATCHING METHOD OF WATER RESOURCES

Optimal principles of water resources dispatch

Optimal allocation of water resources refers to the reasonable allocation of available water resources between regions within a specific region, and after determining the water demand of the water consumption area and existing projects, the utilization rate of limited water resources will be increased (Fang *et al.* 2018).

The optimal scheduling of water resources needs to select the optimal scheduling decision method that meets the system standard functions and constraints to form a model. Moreover, the core of water resources optimization is water regulation. The steps of water resources optimization scheduling are as follows (Luo *et al.* 2018).

(1) Clear scheduling goals and constraints.
(2) Establish a model and choose a model solution method.
(3) Analyze the comparison results and optimize the scheduling plan.

Based on an analysis of the balance between water supply and demand, the optimal allocation of water resources needs to understand the current situation of water supply and demand, so as to clarify the overall direction of future water resources. Moreover,, the optimal allocation of water resources is a multi-stage decision-making problem, which requires to allocate water resources rationally in time, space and quantity, so that limited water resources can produce better comprehensive benefits to achieve the purpose of sustainable development (Liu *et al.* 2018).

Construction and solution of optimal configuration model

The optimal allocation model of water resources system can adopt the method of constructing system constraints and objective functions, and the model is solved to obtain better benefits under the given requirements. Besides, in order to construct the optimal water resources dispatch model for the water station and reservoir system of the canal, the objectives of models need to be clarified. What is more (Gao *et al.* 2018), firstly, for the water station system of the reservoir, the optimization goal is to reduce the system operating cost on the premise of satisfying various water demands to the greatest extent. Secondly, according to the characteristics of regional water resources and reservoir operation rules, the constraints of the model is listed. Then, an appropriate method is chosen to solve the model and draw a program flow chart of the method. Finally, the program is written in the appropriate programming language (Lin *et al.* 2016).

Generalization of water supply system

The water supply system includes a single reservoir that supplies water to the water supply channel of the water receiving area and a single supplementary channel water station that draws water from the river network. When the amount of water in the reservoir is insufficient to meet the demand of the water receiving area, water can be extracted from the upstream river network to the channel to reduce the water shortage in each stage of the water receiving area (Hu 2016). The generalized water supply system is shown in Figure 1.

Among them, L is the amount of water coming from the reservoir, and Z is the amount of evaporative leakage of the reservoir. Besides, Q refers to the amount of water discharged from the reservoir. Additionally, G_n indicates the amount of water supplied by the reservoir, and n_1 and n_2 are the water demand of each water-using area.

Model establishment

Aiming at the above generalized system, a single reservoir and a single external drainage water supply station are taken as the research object (Liu *et al.* 2018). What is more, the minimum sum of squares of water shortage in each period in the system is the objective function, and the decision variables include the water supply volume G_n of the reservoir in each period and the water lift volume P_i of the canal-filling station. Moreover, constraint conditions are composed of the annual water supply of the reservoir, the annual diversion of external water diversion which are water rights and the constraints of the reservoir operation rules. Meanwhile, a mathematical model is established for the optimal operation of water resources in the water supply station and reservoir system (Ha *et al.* 2017), as shown in Figure 2.

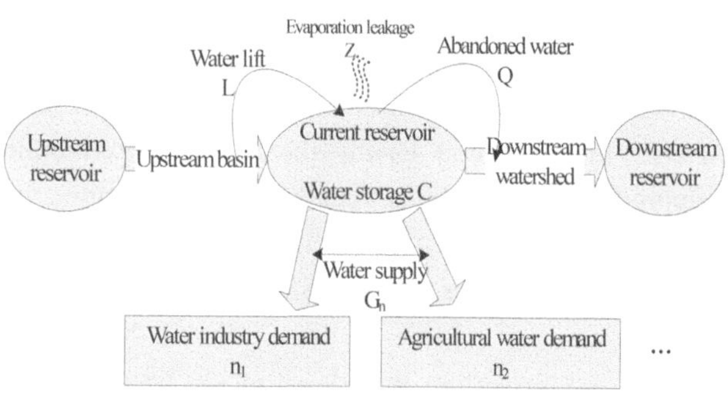

Figure 1 | Water supply channel in water basin.

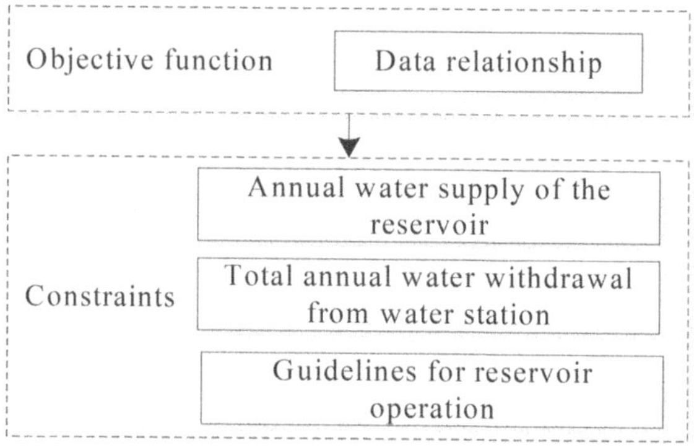

Figure 2 | Mathematical model of optimal water resources dispatch.

(1) Objective function:

$$\varepsilon = \min \tau = \min \sum_{i=1}^{n} (G_{n1} + P_{i1})^2 - PS_{i1}; \tag{1}$$

In formula (1), ε refers to the minimum square sum of the difference between the water demand in the water receiving area and the total system water supply at each time period, and τ is the sum of the square difference between the water demand in the water receiving area at each time period and the total water supply 10^4m^3 of the system. What is more, n is the total number of time periods divided during the year, and i refers to the time period number. Besides, G_{n1}represents the water demand of the water receiving area in the *n1-th* time period, and P_{i1} indicates the water supply volume 10^4m^3/time of the reservoir. In addition, PS_{i1} is the diversion water volume 10^4m^3/time of the repair channel station (Wang *et al.* 2018).

(2) Restrictions

① Restriction on the annual water availability of the reservoir means under different incoming water frequencies, the amount of water the reservoir can provide, namely

$$\sum_{i=1}^{n} G_n \le \varphi \tag{2}$$

In formula (2), φ refers to the total annual water supply of the reservoir10^4m^3.

② The total annual water withdrawal (water rights) of the water station means under different incoming water frequencies, the amount of water that the canal water station can provide (Xiong *et al.* 2017). It is formulated as

$$\sum_{i=1}^{n} G_n \leq \delta \quad (3)$$

In formula (3), δ is the total allowable annual water extraction of the water extraction station 10^4m^3.

③ Restriction of reservoir operation rules are that the storage capacity of the reservoir at the end of each period should not be less than the minimum storage capacity and greater than the corresponding storage capacity of the flood control limit water level (Shao *et al.* 2017), that is

$$\xi_{\min} \leq \xi_i \leq \xi_p \quad (4)$$

In formula (4), $\xi_{\min}$ is the minimum storage capacity 10^4m^3, and ξ_p refers to the storage capacity 10^4m^3 corresponding to the flood control limit of the reservoir.

Among them, ξ_i is related to the storage capacity of the reservoir in the previous period, the amount of incoming water from the reservoir, the amount of discarded water, the amount of water supply and the amount of evaporation and leakage, namely:

$$\xi_i = \xi_{i-1} + Q_{i-1} - P_{i-1} - PS_{i1-1} \quad (5)$$

In formula (5), each variable is the water quantity variable at the end of the i-1 period.

Therefore, when $\xi_i - \xi_p$, $Q_{i-1} = \xi_i - \xi_p$.

THE ESTABLISHMENT AND SOLUTION OF THE OPTIMAL ALLOCATION MODEL OF WATER RESOURCES

Decision variables

The decision variables for the rational allocation of water resources are:

$$\lambda = (\lambda_{ijk}) \quad (6)$$

In Equation (6), $\lambda_i jk$ is the amount of water allocated to the *k-th* water sector by the *j-th* water source in the *i-th* zone, and i refers to the zone number, followed by $(i_1, i_2 \cdots, i_n)$. Moreover, j is the water source number, mainly including weir diversion, diversion of runoff diversion, local reservoir storage and groundwater. In addition, k is the water department number, mainly industrial water, agricultural water, domestic water and ecological water included (Xing 2017).

Objective function

Objective 1: Net benefit of regional water supply is the largest.

$$\max \eta_1(\lambda) = \sum_{i=1}^{6}\sum_{j=1}^{4}\sum_{k=1}^{4} x_{ijk}\lambda_{ijk} \quad (7)$$

In Equation (7), η_1 refers to the maximum benefit, and x_{ijk} is the water supply benefit coefficient of the j water sources in the i zone to the k water sector.

Objective 2: The region's overall water shortage is minimal.

$$\min \eta_2(\lambda) = \sum_{i=1}^{6}\sum_{k=1}^{4}\left(y_{ijk} - \sum_{j=1}^{4}\lambda_{ijk}\right) \quad (8)$$

In Equation (8), y_{ijk} is the total water demand of the i-part k water-using department.

Objective 3: The area has the least amount of groundwater extraction.

$$\min \eta_2(\lambda) = \sum_{i=1}^{6}\sum_{k=1}^{4}\lambda_{ik4} \quad (9)$$

Restrictions

1) Water resource constraints. The formula is as follows.

Water supply restriction at the head of the weir:

$$\begin{cases} \sum_{i=1}^{6}\sum_{k=1}^{4}\lambda_{ijk_1} \leq \theta_{t1}, \theta_{t1} \neq 0 \\ \sum_{i=1}^{6}\sum_{k=1}^{4}\lambda_{ijk_2} \leq \theta_{t2}, \theta_{t2} \neq 0 \\ \sum_{i=1}^{6}\sum_{k=1}^{4}\lambda_{ijk_3} \leq \theta_{t3}, \theta_{t3} \neq 0 \\ \sum_{i=1}^{6}\sum_{k=1}^{4}\lambda_{ijk_4} \leq \theta_{t4}, \theta_{t4} \neq 0 \end{cases} \quad (10)$$

θ_{t1}, θ_{t2}, θ_{t3} and θ_{t4} respectively represent the head of the canal i, the cross-boundary runoff, the local reservoir, and the available water of groundwater. t is the constraint condition.

(1) Constraints on domestic water demand. The formula is as follows.

$$\sum_{j=1}^{4} \lambda_{ijk3} \geq \beta_i \tag{11}$$

In Equation (11), β_i is the domestic water demand in the i-th zone.

(2) The design flow restriction of the branch channel. The formula is as follows.

$$\sum_{k=1}^{4} (\lambda_{ijk1} + \lambda_{ijk2}) \leq \theta_{t5} \cdot t_i, \quad i = 1, 2, \cdots, n \tag{12}$$

In Equation (12), θ_{t5} is the design flow of branch i, m^3/s t_i refers to the number of diversion days of branch i.

(3) Variable non-negative constraints.

$$\lambda_{ijk} \geq 0, i = 1, 2, \cdots, n \quad j = 1, 2, \cdots, n \quad k = 1, 2, \cdots, n \tag{13}$$

Model solution

The steps of the model solving algorithm are as follows.

(1) Real number coding. Real number coding is used in the paper to directly form the individual with the original variables.

(2) Population initialization. The population size is taken as N, and individuals are randomly generated to form a set, while ensuring that it meets the constraints corresponding to the objective function.

(3) Fitness function construction. During the algorithm calculation, each sub-population independently calculates its fitness to determine outstanding individuals. The fitness calculation function is as follows.

$$\mu_i = x \cdot \eta_i + \beta \tag{14}$$

When $\eta_{\min i} \geq \sqrt{\eta_{t6} - \eta_{t7}}$, $x = \dfrac{\eta_{t6}}{\eta_{t7} - \eta_{t8}}$, $\beta = \dfrac{\eta_{t6}(\eta_{t7} - 2\eta_{t6})}{\eta_{t7} - \eta_{t6}}$

When $\eta_{\min i} \leq \sqrt{\eta_{t6} - \eta_{t7}}$, $x = \dfrac{\eta_{t6}}{\eta_{t6} - \eta_{t8}}$, $\beta = \dfrac{-\eta_{t8} \cdot \eta_{t6}}{\eta_{t6} - \eta_{t8}}$ η_{t6}, η_{t7} and η_{t8} respectively represent the mean value, maximum value, and minimum value of the fitness values of genetic genes.

(4) Multi-group genetic algorithm implementation. Operating parameter settings of multi- group genetic algorithm are shown in Table 1.

(5) Output result. By judging whether the population meets the convergence condition and whether the genetic evolution algebra reaches the maximum genetic algebra, the iterations that meet the conditions are terminated, and the best individual obtained is the optimal solution, that is, the optimized various water sources are the optimal water distribution for each division in each department.

SIMULATION AND EXPERIMENT

Simulation test environment

The paper takes the Pohe Reservior's water resources data in various time periods as the research object, and the time span is from August 1, 2018 to August 30, 2019. The reservoir has a total storage capacity of 195 million cubic meters and has flood control and irrigation functions. The annual average rainfall is 1,002.7 mm, and the difference between the maximum and minimum annual rainfall is 1,209.8rmn. The precipitation in May increased significantly, and the precipitation in the flood season was concentrated. The average precipitation from May to September was about 64.0% of the average annual

Table 1 | Multi- group genetic project

Genetic project	Quantity
Subgroup size	70
Single-point crossover probability	0.96
Gaussian mutation probability	0.005
Maximum genetic number	50
Individual replacement percentage	30%
Exchange frequency	4
Number of immigrants	10

precipitation. The reservoir is supplied with water as much as possible to meet the planned water demand. During the optimization period, the reservoir is supplied with water as much as possible to meet the discounted water demand.

(1) The decision variable uses water demand discount coefficient $\psi(\psi \in (0, 1))$ as the decision variable for optimal scheduling. In order to improve the optimization efficiency of the optimization model, the lower limit of ψ is increased from 0 to the minimum water supply intensity of 47.28% in the simulated dispatch. The discounted water requirement of the reservoir is:

$$y_{ijk}^{\psi_i} = \psi_i \cdot y_{jk}^{i} \tag{15}$$

In formula (15), $y_{ijk}^{\psi_i}$ refers to the discounted water demand in the optimization period i, and y_{jk}^{i} is the planned water demand in the optimization period i. Besides, ψ_i is the discount coefficient in the optimization period i, and $\psi_i \in (47.28\%, 100\%)$.

(2) Objective function

In order to solve the problem of low water supply intensity in individual periods during simulated scheduling, and improve the overall water supply during the optimization period, two optimization scheduling goals are set in the model.

Objective a: The minimum value of the water supply intensity in each optimization period is the largest, and the function P_{i1} is:

$$P_{tl} = \sqrt{\max(\min \pi_i)},\ i = 1, 2 \cdots, n \tag{16}$$

In formula (16), π is the maximum water supply intensity of the optimized period i.

Objective b: The average value of water supply intensity in each optimization period is the largest, the function P_{i2} is:

$$P_{i2} = \max \sqrt{\frac{\sum_{i=1}^{n} \pi_i}{n}} \tag{17}$$

In formula (17), i is the optimization period.

There is a certain contradiction among the available storage capacity in the adjacent optimization period, and the planned water demand downstream of the reservoir in the adjacent period is also different, resulting in an inverse proportional relationship between the minimum and average water supply intensity in the optimization period. Therefore, the setting of the two optimization goals is reasonable and effective.

(3) Constraints

The water balance equation and reservoir constraints of the optimized model are consistent with the simulation model. In addition, the decision variable constraint $47.28\% \le \psi_i \le 100\%$ should be satisfied, which is the discount coefficient of water demand.

To sum up, the paper uses the fitting function and improves the genetic gene instead of the penalty function to prevent the penalty solution from exceeding the constraints given by the system, thereby reducing the randomness of the classical genetic algorithm and increasing robustness.

Analysis of simulation scheduling results

Comparison of target extreme value

The traditional linear binary algorithm and the improved genetic algorithm are adopted to solve the dual-objective optimal dispatch model of reservoir water supply. The population size is 100, and the maximum evolutionary generation 2000 is used as the termination condition of the algorithm. What is more, the optimization calculations are independently performed 10 times. Table 2 shows the average optimization results and average calculation time of the two algorithms for each target.

Conclusions are drawn from Table 2.

(1) Judging from the average optimization results of each target extreme value, the average optimization results

Table 2 | Average optimization results and computing time

	Goal 1		Goal 1		Average calculation
Algorithm	**Min/%**	**Max/%**	**Min/%**	**Max/%**	**time/min**
Improved genetic algorithm	64.81	75.68	87.65	88.69	45.55
Traditional linear binary algorithm	58.55	66.87	81.56	83.73	52.39

of the improved genetic algorithm are better than those of the traditional linear binary algorithm.

(2) From the perspective of average calculation time, the average calculation time of the improved genetic algorithm is 19.41% lower than that of the traditional linear binary algorithm.

Analysis of solutions with similar values for single targets shows that the minimum difference between goal 1 and goal 2 is lower than the average, which also reflects that the inherited genes of genetic factors in this algorithm are significantly higher than that of standard genetic algorithms.

Traditional optimization algorithms cannot solve the model reasonably because of premature convergence and exceeding the constraints given by the model. In this paper, a multi-group genetic algorithm based on real number coding is used to solve it through real-coded multi-group genetic algorithm to solve the model solving problem. The optimal configuration results obtained verify the effectiveness of this algorithm.

Comparison of multiple iteration calculations

The best one is respectively selected from the 50 results of reservoir optimal operation in two algorithms, as shown in Figure 3.

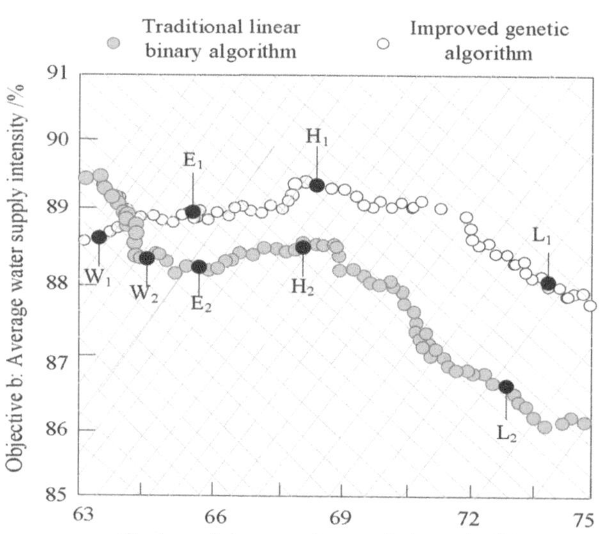

Figure 3 | Comparison of the best optimization results.

The following conclusions can be seen from Figure 3.

(1) For the extreme value analysis of the two targets, W_1 is higher than the corresponding value of W_2 in the two target extreme values, and E_1 is higher than the corresponding value of E_2 in the extreme value of target *b*. However, due to the inverse proportional relationship between goal *a* and goal *b*, it is acceptable that the corresponding value of E_1 is lower than E_2 at the extreme value of goal *a*.

(2) For the analysis of solutions with similar single target values, taking W_2, H_1 and H_2 as examples, on target *b*, H_1 and W_2 are similar and H_1 is slightly larger than W_2. However, on target *a*, H_1 is 9.22% larger than W_2, and H_1 is significantly better than W_2. Similarly, H_1 is significantly better than H_2.

(3) For the shape analysis of the genetic factors convergence, although the accuracy distribution of the traditional linear binary algorithm is wider, its overall distribution position is lower, and the continuity in the upper left part of the coordinates is severely damaged. While the genetic factor of the improved genetic algorithm still maintains a higher overall distribution position of the gene after multiple iterations, and the continuity is better.

In summary, the genetic factor inheritance of the improved genetic algorithm is better than the calculation accuracy of the traditional linear binary algorithm, and the overall optimal solution set corresponding to the improved genetic algorithm calculation accuracy is also better than the optimal solution set corresponding to the traditional linear binary algorithm.

Simulation scheduling of data in different time periods

It can be seen from Figure 3 that the optimal solution set distribution of the improved genetic algorithm can be analyzed, and the range of goal a reaches 10.95%, which is much larger than 0.90% of the range of goal b. In addition, *L* is a turning point for the convergence of genetic factors in the improved genetic algorithm. Compared with W_1, *L* has a large increase in target a in exchange for a small loss of target *b*. Relative to E_1, *L* has a smaller target loss in exchange for a larger increase in target *b*. Meanwhile, the solution set of the improved genetic algorithm as a whole is better than that of the traditional linear binary algorithm.

Therefore, D, which is the optimal solution set of the improved genetic algorithm, is selected as the basis for formulating the optimal operation plan of reservoir water supply in the paper. In addition, the water supply situation corresponding to the optimal solution L is shown in Table 3, and the water supply situation during the destruction period in simulated dispatch and optimized dispatch is shown in Figure 4.

The following conclusions can be seen from Figure 4 and Table 3.

(1) There are a total of 57 water supply destruction periods in the optimized dispatch, and the corresponding guarantee rate during the water supply period is 94.03%, which is slightly lower than the 96.75% guarantee rate during the water supply period for traditional linear binary algorithm scheduling. It shows that the overall adverse effect of optimized dispatch on the water supply of the reservoir is small.

(2) From the perspective of the minimum water supply intensity, the optimization results of the improved genetic algorithm at each stage are better than the results of the traditional linear binary algorithm. Additionally, the minimum value of the water supply intensity is only 49.13% under the traditional linear binary algorithm scheduling, while the optimal scheduling greatly increases it to 99.99%.

(3) From the maximum value of water supply intensity, the traditional linear binary algorithm scheduling result is only 73.44%, while the optimized scheduling result of the improved genetic algorithm is 99.25%.

(4) From the average of water supply intensity, the traditional linear binary algorithm scheduling result is only 69.34%, while the optimized scheduling result of the improved genetic algorithm is 93.04%.

Table 3 | Simulation scheduling optimization data

			Intensity of water supply during failure period%		
Period	**Situation**	**Cumulative damage**	**Minimum value**	**Maximum**	**Average value**
30 days	Normal	3	99.25	99.99	99.62
60 days	Normal	6	82.38	95.64	91.19
90 days	Urgent	9	75.68	89.39	82.54
120 days	Normal	17	79.55	94.89	89.72

In general, the improved genetic algorithm optimization scheduling improves the minimum and average water supply intensity, which makes the water supply intensity change process more stable, thereby reducing the adverse impact of large fluctuations in water supply intensity on downstream production and life.

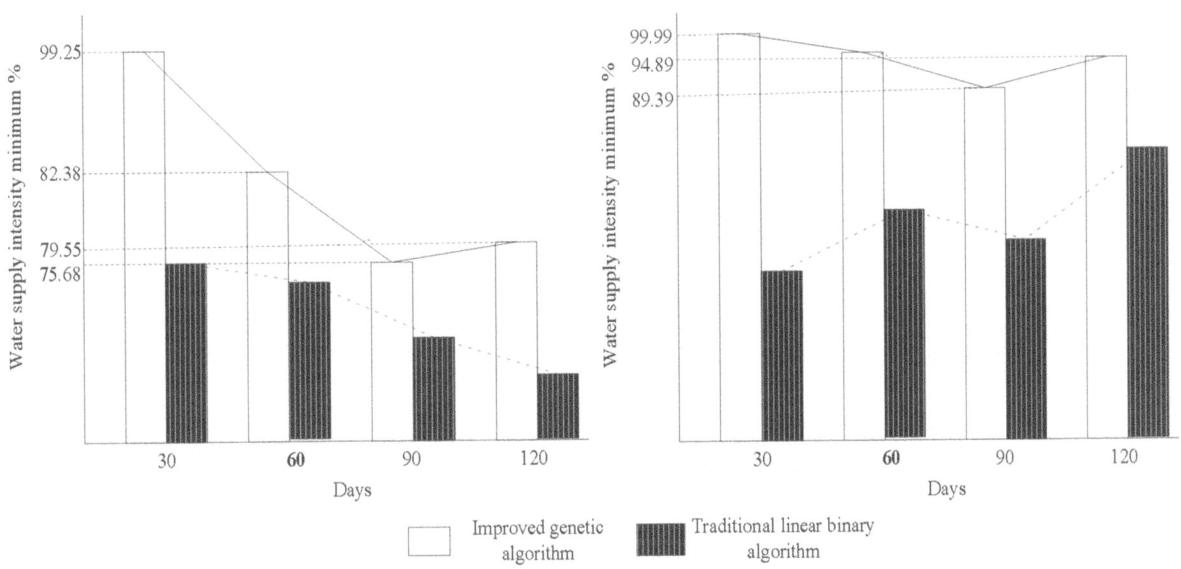

Figure 4 | Water supply during the shortage period.

CONCLUSION

Aiming at the multiple objectives of optimal allocation of water resources, comprehensive consideration should be given to the objectives of optimal allocation of regional water resources such as water supply benefits, water shortage, and groundwater development. Meanwhile, water resource constraints are established. In terms of demand and water supply, a multi-objective water resource optimization configuration model has been established. What is more, the model solution problem is solved by real-coded multiple sets of genetic algorithms, and the obtained optimal configuration results verify the rationality of the built model. Due to the problem of premature convergence in traditional genetic algorithms, a multi-group genetic algorithm based on real number coding is proposed. The use of real number coding effectively avoids the tedious coding and decoding process, can realize that each sub-population searches for the optimal solution at the same time, and prevents the penalty solution from exceeding the constraints given by the system. The algorithm reduces the influence of randomness of classical genetic algorithms and increases robustness. At present, the problem of water conservancy scheduling is limited to the two target mathematical models of data for a certain period of time and the current supply and demand, and the third target model for future demand has not been established. Therefore, how to efficiently solve the multi-objective optimization problem with more than two objectives and combine it with the optimal operation of reservoirs will be the difficulty and focus of future research.

DATA AVAILABILITY STATEMENT

All relevant data are included in the paper or its Supplementary Information.

REFERENCES

Ai, X., Dong, Z. & Mo, M. 2017 Research on multi-objective dispatching model and algorithm of reservoir. *Journal of Hydroelectric Power* **20** (12), 210–222.

Chai, Y., Li, Y., Li, S., Wang, J., Zhu, B. & Chen, M. 2017a Analysis of the trend and cause of water and sediment changes in the main tributaries of the Yellow River Basin since 2000. *Hydropower and Energy Science* **21** (04), 33–50.

Chai, Y., Li, Y., Li, S., Zhu, B. & Wang, J. 2017b Recent changes of water and sediment in the Yangtze River Basin and analysis of their causes. *Journal of Irrigation and Drainage* **12** (03), 23–41.

Fang, G., Ding, Z., Huang, X. & Dai, L. 2018 Study on optimal operation of hydropower station reservoir considering river ecological protection. *Journal of Hydroelectric Power* **07**, 345–356.

Gao, Z., Li, H. & Zhang, Q. 2018 C-D function ridge regression analysis of agricultural water supply benefit in Baojixia irrigation area. *Agricultural Research in Arid Area* **06**, 121–132.

Ha, Y., Bai, T., Huang, Q. & Jin, W. 2017 Research on multi-objective transformation of combined operation of water and sediment in cascade reservoirs. *Journal of Hydroelectric Power* **33** (07), 21–29.

Hu, C. 2016 Development and practice of 'storage, clearing and draining muddy' application method of sandy river reservoirs in my country. *Journal of Water Resources* **24** (03), 54–71.

Lin, L., Zhang, L. & Xu, Y. 2016 Numerical simulation study on hydrodynamics and water quality of river network in Hangzhou Bay New District of Ningbo. *Environmental Science and Management* **41** (06), 45–49.

Liu, Q., Zhong, P., Chen, Y., Xu, B. & Zhu, F. 2018 Improved social sentiment optimization algorithm for cascade hydropower station optimal dispatch. *Journal of Hydroelectric Power* **5** (01), 1320–1333.

Luo, W., Guo, Z. & Kou, X. 2018 Evaluation of Weihe ecological water status and discussion on improvement measures. *Northwest Hydropower* **10** (06), 71–90.

Shao, X., Gu, S., Cao, A., He, H. X. & Yu, T. T. 2017 Wavelet analysis of the characteristics of water and sediment changes in sandy rivers. *Journal of the Yangtze River Academy of Sciences* **11** (05), 109–120.

Tayfur, G. 2017 Modern optimization methods in water resources planning, engineering and management. *Water Resources Management* **31** (10), 3205–3233.

Wang, Z., Wang, X., Lin, J. & Chen, S. 2018 Analysis of water and sediment evolution in Changting section of Tingjiang river basin from 1982 to 2014. *Journal of the Yangtze River Academy of Sciences* **43** (02), 56–71.

Xing, B. 2017 The application of cuckoo search algorithm in the optimal operation of cascade reservoirs. *Hydropower Energy Science* **23** (12), 235–250.

Xiong, H., Chen, X. & Zhang, S. 2017 Method for improving the water quality improvement effect of polluted rivers based on MIKE11 model. *Environmental Science* **38** (12), 5063–5073.

First received 9 July 2020; accepted in revised form 16 November 2020. Available online 30 November 2020

Monitoring of pollution factors of solidified body leachate in a drilling well site and its influence on the surface water environment

Zhikun Liu, Chaoqun Zhang and Qi Li

ABSTRACT

Aiming at the pollution of the leaching solution of the solidified body in the wellbore, a water eutrophication level evaluation algorithm based on the optimized analytic hierarchy process is proposed from the current situation of many factors affecting the eutrophication of the water body and the difficulty of evaluation. Based on this, the user-oriented PC software monitoring system was developed, which mainly includes three major modules: surface water environment monitoring, water eutrophication evaluation, water bloom prediction and warning, and system assistance. The surface water environment monitoring module can receive and display the surface water environment parameter information in real time, and has the functions of data comparison analysis and historical data search, and the eutrophication assessment and water bloom prediction and warning module can analyze the eutrophication status of water bodies in real time and provide short-term and medium-term warnings for water blooms. This research can promote the control of the leachate pollution of the wellbore in the wellbore and the optimization of the surface water environment.

Key words | drilling, pollution factor, surface water environment, well site solidified body

Zhikun Liu (corresponding author)
Chaoqun Zhang
Qi Li
Engineering Research Center of Development and Management for Low to Extra-Low Permeability Oil & Gas Reservoirs in West China,
Ministry of Education,
Shannxi Xi'an 710065,
China
and
Institute of Petroleum Engineering,
Xi'an Shiyou University,
Shannxi Xi'an 710065,
China
E-mail: *gm201803@126.com*

HIGHLIGHTS

- Propose a water eutrophication level evaluation algorithm.
- Develop a user-oriented PC software monitoring system.
- The research can promote the control of the leachate pollution of the wellbore in the wellbore.
- The research can promote the optimization of the surface water environment.
- The water bloom prediction and warning module can analyze the eutrophication status of water bodies in real time.

INTRODUCTION

In recent years, the problem of contamination of the solidified body leachate in surface water environment drilling wells, especially the eutrophication of water bodies, has seriously affected or even threatened industrial and agricultural production and people's lives. Drilling well field solidified body leaching solutions will cause huge pollution of the well site by solidified body leaching of solution in the surface water environment, and monitoring the contamination of the well body solidified body leaching liquid is very important (Wang *et al.* 2017). Compared with the general surface water pollution, the solidified exudate has the characteristics of high salinity, heavy metal content and chemical additives. In surface water pollution monitoring, it is necessary to pay attention to the real-time changes of water salinity,

doi: 10.2166/ws.2020.356

organic matter and other indicators. Therefore, a buoy-type multifunctional surface water environment monitoring system was designed and implemented, and by monitoring the surface water environment information of rivers and lakes, at the same time, the evaluation of water eutrophication and the prediction of water blooms are of great significance to the protection and treatment of the surface water environment (Furukawa *et al.* 2017).

Based on the overall structural framework of the buoy-type multifunctional surface water environment monitoring system, the hardware and software parts of the system have been designed in detail (Kusaiynov *et al.* 2017). In terms of hardware design, the rigid float ball is used as the main body, and the sensing device such as the main controller, water quality sensor, small weather station, and pan/tilt network camera is equipped to obtain water quality parameter information, meteorological parameter information and water surface real-time image (Liu *et al.* 2017). The ARM (Advanced RISC Machines) processor is a low-power microprocessor designed by Acorn Ltd. Among them, based on ARM embedded technology development, the main controller of monitoring, integrating information collection, processing and remote transmission is the focus of hardware design (Sazonov & Mokhov 2017). The program for the main controller in C language, and is a large control system integrating the water quality data collection and processing system, the meteorological data collection and processing system, the GPS (Global Positioning System) geographic information collection and processing system, the SMS alarm sending system and the ARM control system, and it has the characteristics of being able to quickly process a large amount of surface water environment monitoring data, and quickly analyze and process data and transmit data, which is innovative (Li *et al.* 2017). The research is divided into three parts. The first part is the literature review. The second part is the hardware and software development of the pollution factor monitoring of the solidified body leaching solution in the well site, and the software and hardware design. The third part is testing of the system and the proposed method verification.

RELATED WORK

According to Keller and colleagues, there are currently three methods commonly used in water quality information monitoring methods in China: the first is that people collect water samples and then go to the laboratory for analysis, but the monitoring process is cumbersome, and the cycle from the initial collection of water samples to the results is too long (Keller *et al.* 2017). The second is a large-scale fixed water quality automatic monitoring station, Alam and colleagues proposed, its large size, difficult to move, high cost of construction and maintenance (Alam *et al.* 2018). The third is to use a conventional ship to monitor the water body using a portable instrument for the carrier. Xia and colleagues applied it, who and believes that the water quality information obtained is too simple and cannot be transmitted online (Xia *et al.* 2018). Alves and colleagues believe that these three types of monitoring methods have their own shortcomings, and it is impossible to monitor the waters in real time, dynamically and extensively, and it is not effective to carry out water eutrophication evaluation and water bloom prediction and warning (Alves *et al.* 2018). Therefore, Pavlidis & Tsihrintzis applied a buoy-type multipurpose surface water environment monitoring system for rivers and lakes with large water area, which carries out the real-time dynamic monitoring of water bodies. The monitored surface water environment parameters are transmitted to the upper computer software monitoring system, which can monitor the surface water environment parameter information in real time, which has great ecological and social value for protecting surface water environment resources (Pavlidis & Tsihrintzis 2018). Li and co-workers evaluated the eutrophication of water bodies and believes that because of its many influencing factors, the causes are complex and there is no real standard evaluation method (Li *et al.* 2018) Zhongzhi and colleagues believe that evaluators often only pay attention to the influence of surface water environment parameters and ignore the subjective factors such as expert experience. Therefore, evaluation of water body eutrophication has always been an urgent problem to be solved in surface water environment protection work (Zhongzhi *et al.* 2018). Sobrinho and co-workers used an optimized analytic hierarchy process, as well as the surface water environment information and subjective expert experience of the customer view, assessing the degree of eutrophication of the water body, making the evaluation results convincing, thus providing an effective decision-making basis for relevant departments (Sobrinho *et al.* 2018). Although the current research has been rich in the monitoring of surface water environment, there are not many studies on monitoring the pollution factor of the solidified body leachate in the well site and its impact on the surface water environment (Jing 2018). Based on this and the previous research, monitoring of the surface water environment polluted by the solidified body leachate in the well site was studied.

DRILLING WELL FIELD SOLIDIFIED BODY LEACHATE POLLUTION FACTOR MONITORING HARDWARE AND SOFTWARE DEVELOPMENT

The development of the main controller hardware

The main controller of the buoy multifunctional surface water environment monitor is a multifunction master composed of a program burning interface, an Ethernet interface, a serial port interface and the like, which is an ARM architecture, including GPRS (General Packet Radio Service) general packet wireless service module and GPS module. The main controller's CPU (Central Processing Unit) uses a powerful ARM9 series processor with low power consumption, small size and low cost. In addition, compared to the previous series, the ARM9 series selects the Harvard architecture, and its increased pipeline architecture enhances its parallel processing skills; it improved the instruction cycle, which greatly improves the performance of the processor, which can increase performance by about 30%; the main controller developed by this system is shown in Figure 1. In terms of hardware, the main controller mainly includes a central processing unit, a 2-way serial interface, a GPS module, a memory, and a GPRS general packet wireless service module, multiple output voltage, two analog inputs, Ethernet interface, program programming interface, eight dry contact switch inputs and four dry contact switch outputs. Figure 1 shows the main structure of the main controller.

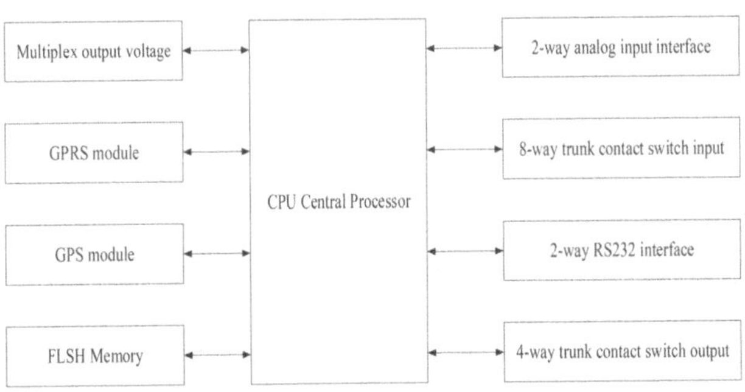

Figure 1 | Main controller structure.

In the main controller of the buoy system, the central processor is connected to eight circuits to control the working process of all functional modules, including data acquisition, analysis, format integration and transmission. As the core of the entire buoy-type multi-purpose surface water environment monitoring system, the central processor controls other functional modules to complete their respective work. In addition to the central processor, the remaining modules in the controller complete their respective functions at the appropriate time. The ferroelectric memory and the FLASH module complement each other and are used to store water quality data information, weather parameter information, and other information. After the RS232 serial port is connected to the RS232-RS485 function chip, it is connected to each sensor, which is used to obtain water quality information and meteorological information of the monitored water body. The GPS module is used for dynamic positioning in monitoring the buoy system, and collects position information of the buoy at various time points. Using the 3G wireless network router, the water quality parameter information collected by the water quality sensor, the meteorological parameter information collected by the small weather station and the geographical location information collected by the GPS are transmitted to the upper computer software by using the 3G wireless network; the GPRS module is used to push the surface water environment information or alarm SMS to the user's mobile phone through the mobile phone network at an appropriate time. In addition to this, the master also has a watchdog circuit and a reset button. The watchdog circuit is used to protect various hardware devices to work properly. The reset button is related to two reset modes. When the reset button is pressed shortly, the hardware device is reset. When the reset button is pressed, the hardware restores the initial value. The network interface circuit is mainly embodied in the transformer circuit and is mainly used to transmit Ethernet signals. In order to ensure the quality of the network signal, the circuit also has anti-interference functions such as signal clutter suppression and impedance matching. The network interface pins are shown in Table 1.

Software function development

The lower part of the buoy system selects the fourth generation version of Code Composer Studio as the lower platform for development. The software is slightly different from other software in that it has a single user interface that can be used to complete every step of the R&D design process. As a powerful software tool that can be used in an integrated development environment, it encapsulates a full set of tools for developing and debugging programs, which integrates editing, compiling, debugging, emulation and other functions to provide convenience for users. The software of the lower part of the buoy system applies the modular design idea, and the different functions are built into different subsystems to speed up the overall running speed of the lower computer. At the same time, it also reduces the amount of work required to upgrade software features. The GPRS general packet wireless service module and the GPS module are simultaneously integrated in the main controller, and the collection, analysis, processing and transmission of water quality information parameters, meteorological information parameters and geographical location information parameters are realized by using various interfaces to connect with various sensors (RS232 interface, analog interface, program programming interface, Ethernet interface, etc.); to send surface water environment information to the user's mobile phone and send a dangerous alarm message to the user's mobile phone; and a remote control function. The main controller applied to the buoy-type multifunctional surface water environment monitoring system can collect surface water environment parameter information (water quality information, weather information, geographical location information) in real time, which has the functions of packaging all kinds of surface water environment data collected by format, then real-time transmission through 3G wireless network, main controller status control function (3G network, temperature), SMS push function and other basic functions. The subordinate machine software part of the buoy system applies the modular design idea to make different functions into different submodules, and the water quality information collection and transmission module, the meteorological information collection and transmission module, the GPS global positioning module, the GPRS/3G wireless network transmission module, and the short message sending module are integrated. The software functional architecture is shown in Figure 2.

The main controller of the buoy-type multifunctional surface water environment monitoring system mainly includes a water quality information acquisition and transmission module, a meteorological information acquisition and transmission module, a GPS module, a wireless network transmission module, and a short message sending module. Among the above modules, the water quality information collection and transmission system module and the meteorological information collection and transmission system module are the cores of the whole software. Due to the role of the timer, according to its own crystal frequency, when the time is up to half an hour, the system will encapsulate the surface water environment data collected by various sensors and GPS into a data packet by parsing the data and arranging the transmitted data format, and finally, the data packet is transmitted to the host computer software monitoring system through the 3G wireless network. The buoy lower machine main program aims to achieve initialization and main body loop. In the main program, the system initialization part includes the central processor,

Table 1 | Network socket pin table

Pin number/name	Pin number/connected chip
9	6-RX −
11	3-RX +
14	2-TX −
16	1-TX +

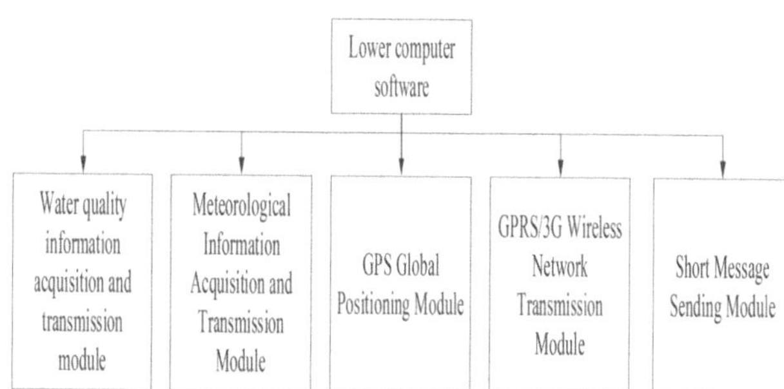

Figure 2 | Software function architecture of lower computer of buoy system master controller.

timer and peripheral chip initialization, as well as parameter configuration in the GPRS module, and obtains the IP address and port number of the upper remote server. The main loop part mainly reads the serial port data expansion around the timing loop. That is, once the number of seconds counted by the timer is reached, the corresponding program is executed. The flow chart of the lower machine main program is shown in Figure 3.

Once the buoy-type multifunctional surface water environment monitor is powered on, the system configuration is first initialized, and the corresponding parameters of the GPRS module are configured in the form of an AT command. After that, it is connected with the address and port number of the upper remote monitoring system, and establishes a TCP transmission control protocol connection with it to prepare for transmitting the surface water environment information to the upper system. After the system initialization process ends, the water quality information, weather information, and geographic location information are collected, processed, and transmitted as the timer expires. In the water quality information collection, processing and transmission module, one of the two RS232 interfaces is used to connect with the water quality sensor, and the water quality information is transmitted to the upper monitoring system by using the 3G wireless network. The flow chart of water quality information collection, processing and transmission module is shown in Figure 4.

When the timing of the timer design arrives, the water quality sensor sends the water quality information collected by the water quality sensor to the main controller, that is,

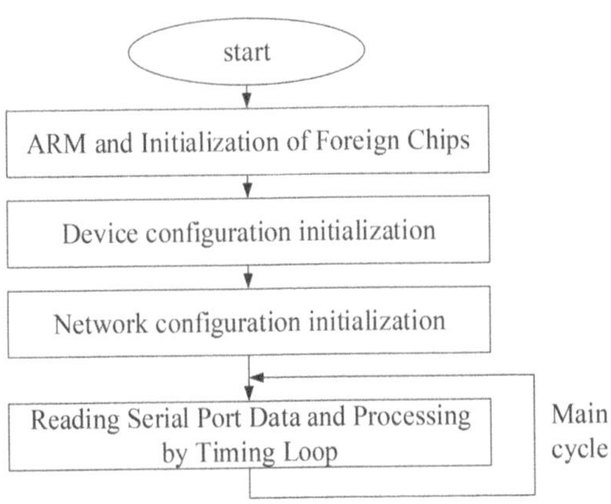

Figure 3 | The main program diagram of the lower computer.

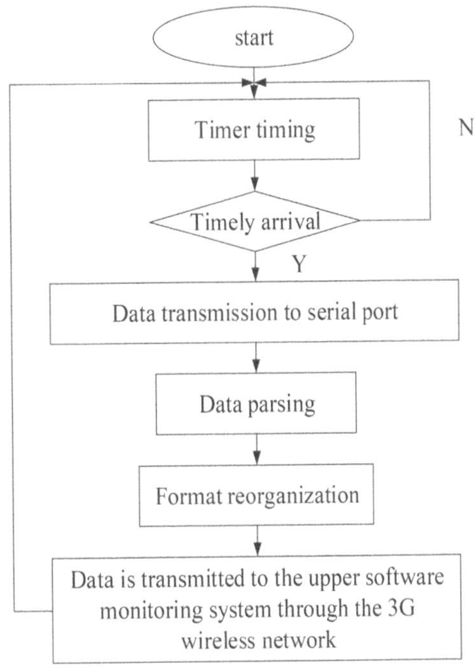

Figure 4 | Flow chart of water quality information collection, processing and transmission module.

the original water quality parameter data. After that, the original data and the recombined data format are parsed in order according to the program in the controller, and the data packet is prepared according to the communication protocol of the upper system. Finally, there is a wait for it to be transmitted to the upper software monitoring system through the 3G wireless network, and complete a cycle, and repeat the process. In this module, another pathway of two RS232 interfaces is connected with a small weather station (wind speed and direction sensor, rain sensor, and illuminance sensor), and the 3G wireless network is used to complete the weather monitoring information sent to the upper monitoring system. The flow chart of the meteorological information acquisition, processing and transmission module is shown in Figure 5.

When the timing of the timer design arrives, the small weather station (wind speed direction sensor, rainfall sensor, illuminance sensor) sends its own collected meteorological information, i.e., the original weather parameter data, to the main controller. After that, the original data and the recombined data format are parsed in order according to the program in the controller, and the data packet is prepared according to the communication protocol of the

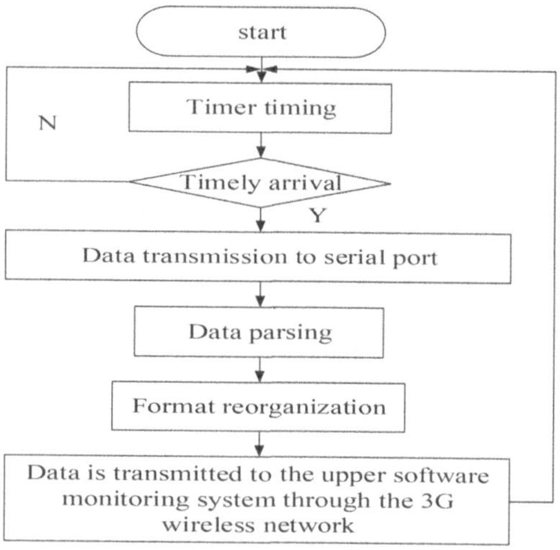

Figure 5 | Flow chart of meteorological information acquisition, processing and transmission module.

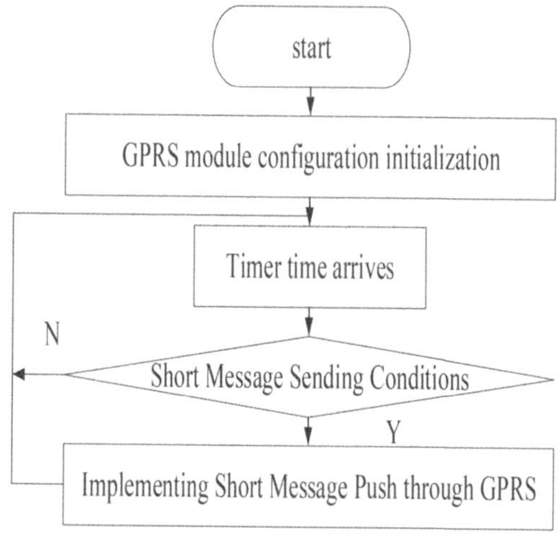

Figure 6 | Short message sending module stream.

upper system. Finally, there is a wait for it to be transmitted to the upper software monitoring system through the 3G wireless network, and complete a cycle, and repeat the process. The GPS module equipped with the GPS receiving circuit and the GPS antenna is integrally packaged into the main controller, and the communication satellite is used to complete the specific positioning of the buoy. The main purpose is to acquire the satellite signal in real time, to accurately track and locate the buoy system, and then determine the geographical location information of the monitoring point, that is, the latitude and longitude coordinates. The work of the short message sending module is mainly completed by means of a GPRS module encapsulated inside the main controller. It is based on the AT command transparent transmission technology, and by inserting a SIM card into the GPRS module, functions such as transmitting information and setting a super user can be completed. The SMS module flow chart is shown in Figure 6.

EXPERIMENTAL DESIGN AND ANALYSIS

Through the software development and hardware design of the main controller, according to the set principle of the upper and lower communication protocols, the computer centralizes control and implementation of the buoy monitoring system. Through the laboratory simulation site, the 3G wireless router is used to form a wireless network, and the main controller is provided to perform surface water environment information transmission to the upper software monitoring system. Through each laboratory test, any problem is continuously displayed in the results, and then the corresponding point improvement is found in the program, and then tested. This cycle is carried out, until the upper software monitoring system can stably receive the surface water environment monitoring information transmitted to the lower computer. Before the surface water environment data are sent, if the 3G wireless network transmission modules and the upper software monitoring system are successfully connected, the word 'NetOK' is displayed on the serial port debugging assistant, indicating that the network connection is completed, then the lower computer starts to transmit surface water environment data. When the upper software monitoring system receives the data packet and unpacks successfully, it will notify the lower computer by returning a status information, and the lower computer ends the thread. When the upper software monitoring system has not successfully connected to the main controller for a long time, the word 'TimeOut' is displayed on the display serial debugging assistant, indicating that the network connection fails, and the operator is reminded to check the network connection.

VC ++ is an object-oriented visual integration programming environment developed by Microsoft. It is widely used because it contains auto-complete and error-removal features. For example, it allows the user to remotely debug the program and step through, and enables the operator to compile the modified program again when debugging. In this process, it is not necessary to re-open the code for debugging at that time. These advantages greatly reduce the time it takes to edit, compile, and link, and are even more apparent on large system software. Most of the time, applications use the 'socket' to make requests to the network or to respond to network requests. There are three steps of server monitoring, client request and connection confirmation from first to last. When the server listens, the upper server and the lower computer are prepared for the connection, waiting for the connection, and in this process, the state of the network is also monitored. When the client requests, the lower computer SOCKET sends a request to connect to the upper server SOKET. Before this, it needs to know the address and port number of the server. Otherwise, the application cannot be made. When the connection confirmation is confirmed, when the upper server responds to the lower application, it will send the description of its SOCKET to the lower computer through the thread, and the lower digit confirms that the two are connected. After that, the upper server replies to the initial server listening state, waiting for other lower computers to issue an application.

GIS technology. GIS is a tool on a computer that can analyze and process the situation of space. In other words, it transforms all the conditions of the Earth, including all events produced, into the form of a graph and then parses it. GIS technology combines the visual experience of the map and the geographic analysis with the functions of some databases, such as search and management. The user can access the upper software monitoring system to observe the real-time water circulation within the different monitoring ranges of the monitored water body. The acquisition of parameter information in surface water environment monitoring and water bloom prediction software mainly includes that the monitored surface water environmental parameters are visualized and that the water quality information parameters obtained by the water quality sensor and the meteorological information parameters obtained by the small weather station (wind speed and direction sensor, rainfall sensor, and illuminance sensor) are provided, including total salt content, dissolved oxygen saturation, temperature, chlorophyll, pH, dissolved oxygen, salinity, redox potential, electrical conductivity, blue-green algae, ammonia nitrogen and other water quality indicators; meteorological indicators such as wind speed, wind direction, rainfall, and illumination. After collecting and processing the data uploaded by each sensor, the main controller of the buoy system transmits the surface water environment information collected on the site to the upper computer software monitoring system through the 3G wireless network transmission module, and after using the network to obtain the corresponding parameters, it begins further analysis. The surface water environment information collected by various sensors is stored in the SQLSever database management system through the upper software monitoring system platform, so that the operator can perform operations such as query and report printing. Based on the communication protocol of the buoy system, the upper system adopts the following program design, and the specific implementation flow chart of data reception is shown in Figure 7.

The upper monitoring software system can receive the data packets transmitted by the lower nodes to the upper level, obtain the correct information through verification,

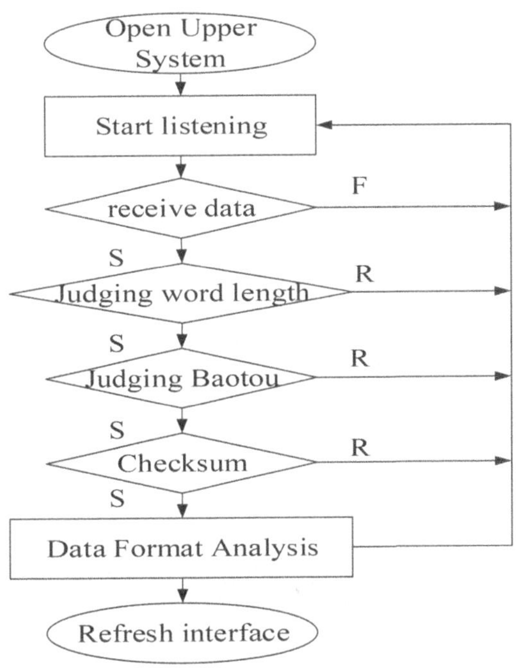

Figure 7 | Data receiving implementation flow chart.

and store the monitoring data in the database. The lower system ensures that the data format is correct, and the on-site water quality information, meteorological information and location can be transmitted to the upper position, and monitored by the upper staff. The upper monitoring software uses the 3G wireless network to receive the surface water environment information sent by the master controller. After that, it analyzes the surface water environment parameters and then starts intelligent integrated management and control, thus completing the functions of real-time information, historical information search, and water eutrophication assessment. In this way, the monitored remote water condition and data analysis processing can be better viewed by the upper remote server. The figure above shows the status of the surface water environment information transmitted by the upper software monitoring system to the lower computer. The data information is displayed in the box below the software interface. If the surface water is polluted by the solidified leaching solution of drilling well site, the organic matter content in the surface water will increase, and the corresponding water quality indexes such as cod will reflect obvious abnormality. In case of surface water pollution, when the pollution exceeds a certain range, water eutrophication will occur. Table 2 shows the monitoring results of surface water environment in the software system. According to the monitoring results in Table 2, there is no surface water pollution problem caused by solidified body leaching solution in the current monitoring sites.

The right side of the software screen shows the real-time connection status of the network. The operator can judge whether the upper software monitors whether the system is connected to the controller by whether the thread is connected or not. Through the linkage test of the upper and lower positions, it confirms that the water quality information, meteorological information and geographical location information in the buoy system are transmitted

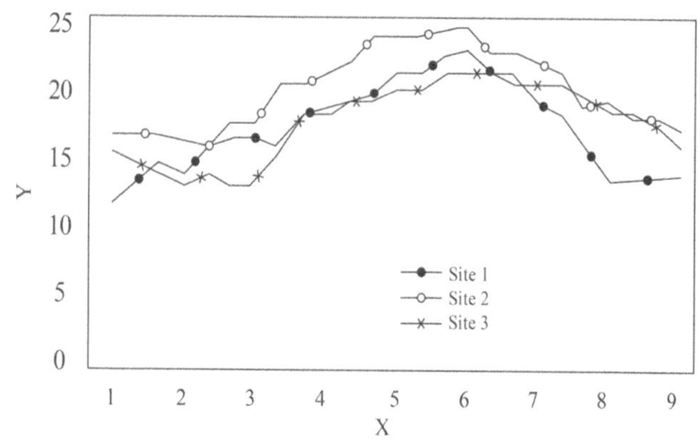

Figure 8 | Data analysis community.

stably, and the functional requirements are realized. The data analysis function is also in the toolbar above the main interface of the software, by determining the start and end dates, specific parameters, and site numbers of the information being queried, the detailed range of changes of the required parameters in the set time period can be displayed in the interface. Furthermore, by selecting the comparison site, the range of changes of the same parameters of different sites in the same time can be observed on the interface, and the monitored water body is comprehensively evaluated, as shown in Figure 8.

CONCLUSION

A buoy-type multifunctional surface water environment monitoring system based on embedded technology is designed, and an optimized analytic hierarchy process is proposed to determine the weighting algorithm for water eutrophication evaluation. The integrated design of the hardware part of the buoy system is completed, including the hierarchical construction of the buoy, the integration of various sensors and instruments, and the design of the main controller. The ARM9 series central processing unit

Table 2 | Statistics of surface water environment monitoring data

Sampling location	pH	Chemical oxygen demand	Ammonia nitrogen	Petroleum	Chloride	Hexavalent chromium
River 170 m to the east of the well pad	7.78	12	0.054	0	37.3	0
Ditch 5 m below cuttings pool	7.84	13	0.063	0	39.3	0
Standard value	6–9	≤20	≤1.0	≤0.05	/	≤0.05

is applied to the main controller, and the collection, processing and transmission functions of the surface water environment information (water quality information, weather information, geographical location information) and the SMS push function are realized according to the preset function requirements. Different from other buoys, the addition of the PTZ network camera enables the buoy to monitor the water surface in real time for 24 hours, and supports functions such as video playback and mobile phone viewing of water surface images. The application of the upper software monitoring system realizes real-time receiving and displaying the surface water environment monitoring information sent by the main controller, historical data query, data comparison analysis and other surface water environment monitoring functions. In addition, the application software can also implement system-assisted functions such as user login, report printing, and network status viewing. The study provides a theoretical reference for the monitoring of the pollution factor of the solidified body leachate in the well site and its impact on the surface water environment. In the next step, the networking mode can be increased, and other networking methods can be tried, such as receiving data using the mobile app, to further improve work efficiency.

DATA AVAILABILITY STATEMENT

All relevant data are included in the paper or its Supplementary Information.

REFERENCES

Alam, L., Mokhtar, M. & Alam, M. M. 2018 Seasonal variation and preliminary risk assessment of trace element pollution in surface water from Langat River, Malaysia. *Social Science Electronic Publishing* **10** (1), 19–40.

Alves, D. D., Riegel, R. P. & De, D. Q. 2018 Seasonal assessment and apportionment of surface water pollution using multivariate statistical methods: Sinos River, southern Brazil. *Environmental Monitoring & Assessment* **190** (7), 384.

Furukawa, Y., Mukai, K. & Ohmura, K. 2017 Improved slant drilling well for in situ remediation of groundwater and soil at contaminated sites. *Environmental Science & Pollution Research* **24** (7), 1–8.

Jing, Z. 2018 Adaptive PI control of bottom hole pressure during oil well drilling. *IFAC-PapersOnLine* **51** (4), 166–171.

Keller, D. H., Horwitz, R. J. & Mead, J. V. 2017 Natural gas drilling in the Marcellus Shale region: well pad densities and aquatic communities. *Hydrobiologia* **795** (1), 1–16.

Kusaiynov, K., Nussupbekov, B. R. & Shuyushbayeva, N. N. 2017 On electric-pulse well drilling and breaking of solids. *Technical Physics* **62** (6), 867–870.

Li, J., Miao, H. & Hui, Z. 2017 Numerical model for evaluating well kick and lost circulation during managed pressure drilling. *Chemistry & Technology of Fuels & Oils* **53** (4), 557–568.

Li, G. Z., Liu, G. R. & Wei, L. 2018 Analysis and optimization of control algorithms for RSS TSP for horizontal well drilling. *Journal of Petroleum Exploration & Production Technology* **8** (4), 1–10.

Liu, H., Wang, B. & Zhao, L. 2017 Drilling fluid technology for use in drilling well sy125 in Syl Gas field in country B. *Drilling Fluid & Completion Fluid* **34** (3), 44–48.

Pavlidis, G. & Tsihrintzis, V. A. 2018 Environmental benefits and control of pollution to surface water and groundwater by agroforestry systems: a review. *Water Resources Management* **32** (1), 1–29.

Sazonov, Y. A. & Mokhov, M. A. 2017 Development of hybrid equipment for well drilling and hydrocarbon production from offshore deposits. *Chemical & Petroleum Engineering* **53** (10), 1–5.

Sobrinho, N. M. B. D. A., Ceddia, M. B. & Zonta, E. 2018 Spatial variability and solubility of barium in a petroleum well-drilling waste disposal area. *Environmental Monitoring & Assessment* **190** (4), 228.

Wang, H., Ge, Y. & Lin, S. 2017 Technologies in deep and ultra-deep well drilling: present status, challenges and future trend in the 13th five-year plan period (2016–2020). *Natural Gas Industry* **4** (5), 319–326.

Xia, J., Wang, L. & Yu, J. 2018 Impact of environmental factors on water quality at multiple spatial scales and its spatial variation in Huai River Basin, China. *Science China Earth Sciences* **61** (1), 1–11.

Zhongzhi, H., Zhanghua, L. & Zaiming, W. 2018 Improving the lubricity of extended reach, well. drilling equipment using non-fluorescent extreme pressure cationic lubricants. *Chemistry and Technology of Fuels and Oils* **54** (4), 513–522.

First received 29 July 2020; accepted in revised form 15 November 2020. Available online 7 December 2020

Virtual water and its effect in Huaihe River basin: an analysis using a dynamic multi-region computable general equilibrium model

Jing Zhao, Jinhua Liu, Hongzhen Ni, Yong Zhao, Yuping Han and Zicheng Diao

ABSTRACT

A dynamic multi-region computable general equilibrium (CGE) model is used to analyze virtual water and its effect in the Huaihe River basin, which includes Jiangsu, Anhui, Shandong and Henan provinces. The baseline scenario was first constructed to describe the trade situation and virtual water among different regions and sectors from 2009–2020. Then two hypothetical policy scenarios were set to analyze the economic, water consumption and virtual water trade impacts under different policy shocks. The results show that (1) the Huaihe River basin was a net virtual water outflow region, with 13.977×10^9 m^3 virtual water transferred outside the river basin in 2009, and 26.653×10^9 m^3 VW transferred to outside in 2020. Jiangsu is the largest virtual water outflow province (mainly flows to outside), followed by Anhui. Anhui suffers from severe water shortage, but provides a net virtual water outflow to Henan, Jiangsu and Shandong and outside; (2) reducing Anhui's agricultural output (S1 scenario) and Anhui's agricultural virtual water outflow (S2 scenario) can not only reduce virtual water trade but also save water, but S1 scenario reduced net virtual water outflow by more, and saved more water. S1 had a positive effect on gross domestic product (GDP), but reduced the employment rate.

Key words | dynamic multi-region CGE model, Huaihe River basin, macroeconomic impact, virtual water trade, water consumption

Jing Zhao
Yuping Han
Zicheng Diao
College of Water Resources,
North China University of Water Resources and Electric Power,
Zhengzhou 450046,
China

Jinhua Liu
Huadong Engineering Corporation Limited,
Hangzhou 311122,
China

Hongzhen Ni
Yong Zhao (corresponding author)
State Key Laboratory of Simulation and Regulation of Water Cycles in River Basin,
China Institute of Water Resources and Hydropower Research,
Beijing 100038,
China
E-mail: *zhaoyong@iwhr.com*

HIGHLIGHTS

- Role of VWT analyzed using a CGE.
- CGE based on 12 years of data (2009–2020).
- VW study not only focused on agriculture, but also included industry and service sectors.
- Two hypothetical scenarios were set to analyze the economic, water consumption and VWT impact.
- Assessment of VW will change water usage and trade in different upstream and downstream regions of the Huaihe watershed.

INTRODUCTION

Many countries are suffering from water scarcity and it affects a third of the people in the world. Water scarcity is a serious problem in China and has been getting worse in recent years. China started construction of the South-to-North Water Transfer Project in 2002 to alleviate the uneven spatial and temporal distribution of water, and the 'sending grains from the north to the south' policy was formulated and has been carried out since 1990. So China's

doi: 10.2166/ws.2020.373

net virtual water trade (VWT) pattern is roughly the opposite of the distribution of its water resources.

VWT has often been recognized for its ability to improve physical and economic access to water in water-scarce regions, allowing regions to save domestic water through the import of water-intensive products (Kumar 2018). The term 'virtual water' (VT) was first proposed in 1994 by Allan (1994). Virtual water is defined as the water embodied in goods during the production process (Hoekstra 2003). VWT has been much studied in recent years. Some studies have focused on the micro-level, such as the virtual water content of products or consumers; for example, Hoekstra & Chapagain (2008) and Chapagain & Hoekstra (2011) have researched the VW content of coffee, tea, rice and meat. Some studies have focused on the national and regional levels, such as those of Hoekstra & Hung (2005), Duarte *et al.* (2019), Lenzen (2009), Hanasaki *et al.* (2010). They evaluated the national or regional VW balance in relation to water needs and water availability. Other studies have focused on the macro and global levels; for instance, Hoekstra & Hung (2005) estimated the global VW exports related to international trade in crops from 1995 to 1999. Chen & Chen (2013) studied the VW account of the global economy, with VW profiles of 112 nation-level regions. Their findings highlighted the importance of VWT in water-scarce regions.

Recently, VW calculation methods have also attracted much attention. The methods mainly fall into two categories: one is the production tree method (Hoekstra *et al.* 2012), and the other is the input–output (IO) method. The IO model is a top-down method, which uses water use data and input–output tables to calculate the direct water coefficient and total water coefficient for each sector or region (Velázquez 2007). The IO model has been utilized as an effective method to assess resource flows and transfers. Some researchers, such as Wang *et al.* (2013), Arto *et al.* (2016), Shi & Zhan (2015) and Zhang *et al.* (2020) have used the IO model to evaluate VW flows. Other researchers such as Jiang *et al.* (2015) and Deng *et al.* (2015), have used the multi-regional IO analysis method to analyze VWT among various regions. Their studies have verified that the IO model clearly quantifies inter-regional and inter-sectoral VW flows, and represents both direct and indirect water input during production processes. Other studies have used a computable general equilibrium (CGE) model to investigate VWT; for example, Roson & Sartori (2010) applied a CGE to the Mediterranean region to assess to what extent VW is a viable adaptive option for water-scarce regions. Their results suggested that VW could help reduce the impacts of water scarcity, but only marginally. Berrittella *et al.* (2007) also used a CGE model to research the role of VW and water scarcity in the context of international trade. They found that VWT could help balance the supply and demand of water resources in regions suffering from water scarcity. The CGE model provides a systemic and disaggregated data representation of multi-regional and national economies. They fully account for circular income flows, inter-sectoral and market linkages. The CGE model has been used as a means to identify the hidden water consumers along the whole supply chain.

Despite the vast body of literature on VWT, however, (1) VWT research have focused more on agriculture, and the VW of industry and services has hardly been studied; (2) structural changes in the economy will lead VWT changes over time. Most work thus far has focused on assessing VWT over a specific year. The effects of long-term changes in VWT have barely been studied; (3) some studies have used the IO model to compute VWT, but the IO model is a static model that considers a specific year.

To the best of our knowledge, no one has studied VW flows using a dynamic (over time) multi-regional CGE model. This study endeavors to bridge these gaps by using a dynamic CGE model, the SinoTERM model (the enormous regional model), which includes 137 sectors in 31 regions on the Chinese mainland, to analyze VWT and its effect on the Huaihe River basin. The concept of VW was incorporated into the CGE model by adding the VW module to the CGE model. The CGE model helps realize a dynamic simulation through annual capital accumulation and regional labor market dynamic adjustment. The baseline scenario simulation is based on 12 years of data (2009–2020). This paper discusses the trade situation and VW features of the Huaihe River basin from 2009–2020. It sets two policy scenarios (the policy simulation result analysis is from 2019 to 2020), to analyze the economic, water consumption and VWT impact under different policies.

MODELING FRAMEWORK

A dynamically multi-regional CGE model of the Chinese economy (also called SinoTERM) was developed by the Centre of Policy Studies at Victoria University in Australia based on the Australian TERM. The theory of the SinoTERM model is similar to national dynamic CGE models such as MONASH (Dixon & Rimmer 2002) and CHINAGEM (Mai *et al.* 2010) except that it has multiple regions. Each region in the model is treated as a separate economy, linked to the other regions by trade. The equation system of SinoTERM is similar to other models of the TERM family and is described in Wittwer & Horridge (2008).

The features of the SinoTERM model include full input–output databases, inter-regional trade matrices and behavioral equations for each province. The SinoTERM database contains input–output data for 137 sectors in 31 regions of China's mainland. It includes detailed inter-regional trade, consumption, import, and export data. It includes detailed information under 'margins', covering rail, road, water and air freight, plus pipelines, warehousing, trade (retail plus wholesale) and insurance. The database includes detailed tax matrices, allowing for the possibility (either in the original data or as part of a simulation) that tax rates on a particular good might vary both between regions and users. The equations in the model are linearized for simplicity, which, combined with accuracy via multi-step solution methods, ensures model efficiency through the use of GEMPACK software. This model provides a disaggregated representation of national, regional and multi-regional economies, which fully accounts for circular income flows, inter-sectoral and market linkages. The corresponding resource and environmental impacts resulting from consumption activities in one region can be traced to specific production sectors in other regions through the inter-regional supply chain (Robson *et al.* 2018).

Dynamic mechanisms

Dynamic mechanisms have been added to SinoTERM (Wittwer & Horridge 2008). The main enhancement of dynamics was to link capital and investment over time.

(1) Physical capital accumulation

Capital in industry j accumulates according to:

$$K_j(t+1) = K_j(t)*(1 - D_j) + I_j(t) \tag{1}$$

$K_j(t)$ is the quantity of capital available for use in industry j during year t; $I_j(t)$ is the quantity of new capital created for industry j during year t; D_j is the rate of depreciation, treated as a known parameter.

The mechanism for determining j's investment can be represented by:

$$\begin{aligned} E_t[ROR_j(t)] &= -1 + \frac{E_t[Q_j(t+1)]}{C_j(t)} * \frac{1}{1+r} \\ &\quad + (1 - D_j) * \frac{E_t[C_j(t+1)]}{C_j(t)} * \frac{1}{1+r} \end{aligned} \tag{2}$$

$$E_t[ROR_j(t)] = f_{jt}\left(\frac{K_j(t+1)}{K_j(t)} - 1\right) \tag{3}$$

where E_t denotes an expectation held in year t; $ROR_j(t)$ is the rate of return on industry j's investment undertake in year t; $Q_j(t+1)$ is the rental on j's capital in year $t+1$; r is the rate of interest; $C_j(t)$ is the cost of an extra unit of capital installed for industry j in year t; and f_{jt} is a non-decreasing function.

Equation (2) defines the expected rate of return for industry j in year t as the expected present value of an extra dollar of investment. Equation (3) defines an investment – supply curve: it shows how the rate of return that investors require if they are to advance an extra dollar to industry j depends on the rate of growth of j's capital stock.

(2) Regional labor market dynamics are also introduced into the SinoTERM model. This allows partial adjustment in regional labor markets to occur over time in response to an economic shock. The regional labor market adjustment mechanism in SinoTERM, in level form, is given by:

$$\left(\frac{W_t^r}{Wf_t^r} - 1\right) = \left(\frac{W_{t-1}^r}{Wf_{t-1}^r} - 1\right) + \alpha\left(\frac{EMP_t^r}{EMPf_t^r} - \frac{LS_t^r}{LSf_t^r}\right) \tag{4}$$

If the policy shock weakens the labor market in region r at period t relative to forecast, real wages W_t^r in policy scenario will fall relative to forecast Wf_t^r. In addition, there will be an initial enlarged gap between the labor market demand EMP_t^r and supply LS_t^r relative to forecast levels $EMPf_t^r$ and LSf_t^r. In successive years, the gap between demand and supply will gradually return to forecast levels through a further decline in real wages. The speed of labor market adjustment is governed by a positive parameter α.

With the dynamic approach, the long-term effect of VWT is represented year-by-year in a simulation.

Virtual water module

The VW module assigned is the water used by final consumption rather than intermediate consumption, which allows the researcher to assess both direct and indirect water requirements of the final consumption pattern of a nation, region, lifestyle group, or household.

The SinoTERM model includes R provinces and N economic sectors, so the total output (X_j^r) of sector j in province r can be expressed as intermediate input ($A_{ij}^r X_j^r$) plus final consumption (Y_j^r):

$$X_j^r = A_{ij}^r X_j^r + Y_j^r \tag{5}$$

Equation (5) be rewritten as:

$$X_j^r = (I - A_{ij}^r)^{-1} + Y_j^r \tag{6}$$

where I is the identity matrix and has the same order as the matrix A_{ij}^r.

The vector of direct water consumption is extended as: $W_j^r = [W_1^r, W_2^r, \cdots\cdots, W_n^r]$, where W_j^r is the direct water consumption vector of region r.

The water consumption coefficients (water supplied by region r to regions s to generate one monetary unit of final demand) can thus be derived by:

$$D_j^{rs} = (I - A_{ij}^{rs})^{-1} W_j^{rs} \tag{7}$$

where D_j^{rs} denotes the total water supplied by region r to other regions to generate one monetary unit of final demand in sector j of region s.

The virtual water from region s to region r can be calculated as:

$$VW_{ij}^{rs} = D_{ij}^{rs} Y_j^{rs} \tag{8}$$

where VW_{ij}^{rs} is the virtual water volume from i sector in region r to j sector in region s and Y_j^{rs} the output of j sector in region s.

Thus, the total virtual water inflow and outflow of region r can be calculated as:

$$VW_{in}^r = \sum_{i \neq r} VW^{ir} \tag{9}$$

$$VW_{out}^r = \sum_{i \neq r} VW^{ir} \tag{10}$$

where VW_{in}^r is the total virtual water inflow of region r, and VW_{out}^r is the total virtual water outflow of region r.

The SinoTERM model realizes dynamic simulation by capital dynamic accumulation and regional labor market dynamic adjustment. Capital dynamic accumulation has been discussed in detail in Dixon & Rimmer (2002) and Mai *et al.* (2010). Regional labor market dynamics have been discussed in Wittwer & Horridge (2008).

SCENARIO DEVELOPMENT

A baseline was developed first. It is a business-as-usual scenario without any policies. Then a policy simulation was conducted, and a shock with the VWT policy was added. The effects of VWT are measured by deviations of variables in the policy simulation from their baseline levels.

Research area

The Huaihe River basin is located in the eastern part of China and originates from Tongbai Mountain in Henan Province. The river flows eastward for approximately 1,000 km through Henan, Anhui, Jiangsu and Shandong provinces (Figure 1). The annual precipitation ranges from 600 to 1,500 mm. The uneven spatial and temporal distribution of water resources results in water scarcity. The multi-year average natural runoff of the Huaihe River basin occupies only 2% of China, per-capita annual runoff is only 27% of China, and only 1/20 of the world's average. But the Huaihe River basin accounts for 13% of the population, 7.7% of the gross domestic product (GDP), 12% of the cultivated land area, 19% of food

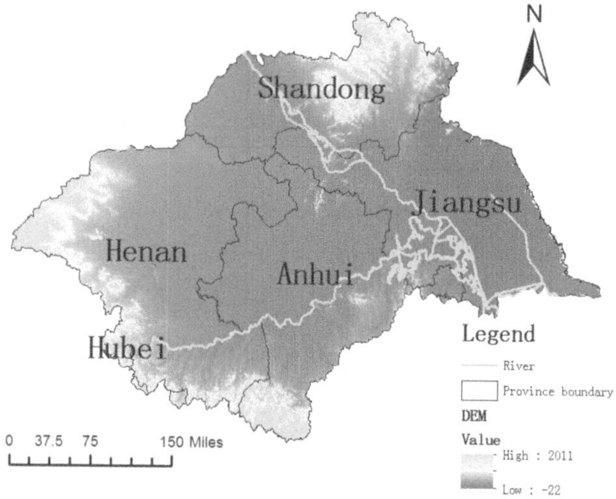

Figure 1 | Map of Huaihe River basin.

production, and 25% of the food that is exported. Rapid economic development and high-density population have exacerbated Huaihe's water scarcity, which restricts the sustainable socio-economic development of the Huaihe River basin.

The four provinces are all water shortage areas, and the amount of water available per capita is far lower than the national average. Jiangsu has the most developed economy in the Huaihe River basin, with a per-capita GDP of 115,168 yuan in 2018, followed by Shandong (per-capita GDP was 76,267 yuan in 2018), Henan (per-capita GDP was 50,152 yuan in 2018), and Anhui (per-capita GDP was 47,712 yuan in 2018). The four provinces in the Huaihe River basin and other regions in the outside of the Huaihe River basin all have VWT, as described in the 'Virtual water trade' section.

Regions and sectors

The database of the SinoTERM model consists of value matrices of commodities, sectors, sources and regions. In order to focus on the VW issue and speed up solution times, we aggregated the sectors and regions to manageable dimensions while retaining detail in the sectors. We also aggregated the SinoTERM database of 137 sectors to 18 sectors. The 18 sectors are listed in Table S1 (Supplementary Information). The Huaihe River basin was divided into four regions: Henan, Anhui, Jiangsu, Shandong.

Baseline scenario

Macro economy

For the baseline scenario, macroeconomic data was divided into two periods, the first (historical) period is from 2009 to 2018 and the second (planned) period is from 2019 to 2020. For the first period, economic variables are assigned values according to the actual economic growth, including main economic indicators such as GDP, consumption, investment, industrial structure, employment and price levels, etc. For the second period, economic data are assigned according to the 13th Five-Year Plan of four provinces. The predictions for economic growth are displayed in Table 1, and the agricultural added value and percentages are displayed in Table 2.

Inter-provincial trade flows

Inter-provincial trade data are obtained from the Inter-regional Input–Output table of 2009. The trade flow of 18

Table 1 | GDP and its speed of growth in the baseline scenario (10^9 yuan, at 2009 prices)

	GDP							GDP growth speed	
	2009	2015	2016	2017	2018	2019	2020	2019	2020
Henan	104	187	202	218	234	249	264	6.50%	5.85%
Anhui	39	63	68	74	80	85	91	6.70%	6.03%
Jiangsu	97	167	180	193	206	219	232	6.50%	5.85%
Shandong	70	124	133	143	152	162	171	6.30%	5.67%
Total	310	540	584	629	673	717	759	6.55%	5.90%

Table 2 | Agricultural added value and percentages in 2009 (10^9 yuan, at 2009 prices)

	Henan	Anhui	Jiangsu	Shandong	Total
Agricultural added value	1.64	0.88	1.31	0.88	4.71
Agricultural added value as a percentage	34.8	18.7	27.8	18.7	100

sectors of four provinces among the Huaihe basin are shown in Table S2 (Supplementary Information). Henan and Anhui are net trade outflow areas. Jiangsu and Shandong are net trade inflow areas. From the perspective of different sectors, Henan's primary inflows are agriculture, coal, food and textile, and outflows are metallurgy and machinery. Anhui's primary inflows are petroleum, chemistry and construction, and the outflow is agriculture. Jiangsu needs inflows of agriculture and food, and its outflows are machinery. Shandong has an obvious advantage in the petroleum and chemistry sectors, but for agriculture and some service sectors it relies on other regions. Table 3 shows the inter-province trade value flow. The total trade value of the Huaihe River basin was $8,886 \times 10^9$ yuan.

Water data

Water data are collected from various sources. The water use by sector is extracted from the water resource bulletin of Huaihe. Only blue water (i.e., fresh surface and groundwater withdrawals) is included in these statistics. Green water (i.e., precipitation stored in the soil) is not included in this study.

To describe the VW content of the commodities, direct and total water use coefficients were computed according to Hoekstra (2003), Hoekstra & Chapagain (2008), and Chapagain & Hoekstra (2011). The direct water use coefficient and total water use coefficient of the Huaihe River basin are shown in Table S3 (Supplementary Information). Agricultural output increased per thousand yuan, and it consume a total of 7,183 m^3 of water, 5,118 m^3 directly and 2,065 m^3 indirectly. The VW content of agriculture is the highest, followed by food and electricity. The total water use coefficient of the same sector between provinces is different.

Table 3 | Inter-province trade flow (10^9 yuan) in 2009

Outflow Inflow	Henan	Anhui	Jiangsu	Shandong	Total
Henan	2,390	85	136	131	2,742
Anhui	82	798	81	41	1,002
Jiangsu	183	117	2,759	96	3,155
Shandong	142	86	114	1,645	1,987
Total	2,797	1,086	3,090	1,913	8,886

Virtual water trade

Figure 2 shows VWT of four provinces. The net outflow of VW from the Huaihe River basin was 13.977×10^9 m^3. The net outflow of agricultural VW was 8,753 m^3, accounting for 63% of the total VW (Table S4, Supplementary Information). Agriculture acounts for most VWT, and often decides the direction of VWT among regions. The industrial sector is an important part of VWT and affect the scale of VWT among regions. The service sector water use coefficient is small and does not have a great impact on VWT patterns. Henan and Anhui province were net VW outflow regions. Jiangsu and Shandong provinces were net VW inflow regions in 2009, and the net inflow volumes were, respectively, 300×10^6 m^3 and 145×10^6 m^3 in the Huaihe River basin. VW outflows from Jiangsu and Anhui, respectively, were $6,204 \times 10^6$ m^3 and $4,188 \times 10^6$ m^3 in 2009. VW outflows from Henan and Shandong, respectively, were $2,387 \times 10^6$ m^3 and $1,198 \times 10^6$ m^3 in 2009. Henan, Anhui and Jiangsu's VW flows to Shandong were 442×10^6 m^3, 468×10^6 m^3 and 535×10^6 m^3 respectively. Anhui was a net outflow region without VW inflow. Anhui's VW outflows were 76×10^6 m^3, 536×10^6 m^3 and 648×10^6 m^3 to Henan, Jiangsu and Shandong, respectively.

The flow of VW in 2020 is shown in Table 4. The Huaihe River basin will transfer $26,653 \times 10^6$ m^3 VW to other regions outside of the basin.

Other parameters

The labor demand elasticity was 0.243, according to the Chinese Academy of Social Sciences (Zheng & Fan 2008). Consumer price elasticity was set at 4, according to the PRCGEM model data of the Chinese Academy of Social Sciences (Zheng & Fan 2008). The Frisch parameter and the Armington elasticity were obtained from the CHINAGEM model (Mai *et al.* 2010). The elasticity of substitution between

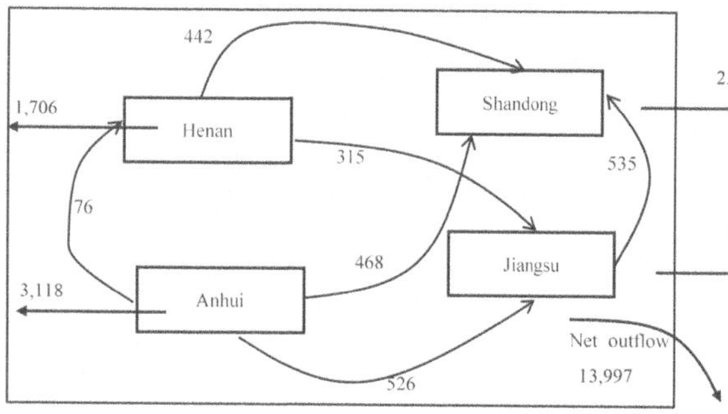

Figure 2 | Pattern of VW flow of the Huaihe River basin in 2009 (10^6 m^3).

primary factors, constant elasticity transformation, the elasticity of household demand expenditure, etc. was obtained from the China version of the ORANI-G model the Centre of Policy Studies at Victoria University (Dixon & Rimmer 2002).

Policy scenario

The net outflow VW region is mainly concentrated in Anhui, which is a water shortage region. In order to alleviate the water stress in the upper reaches of the Huaihe River basin, the outflow of water-intensive commodities should be reduced. Two policy scenarios were set: in S1 Anhui's agricultural output is reduced by 5%; in S2 Anhui's agricultural outflow demand is reduced by 5%.

RESULTS

Macroeconomic impact

The macroeconomic impact under the two policy scenarios is shown in Figure 3. In scenario S1, Anhui's agricultural output is reduced by 5%, so its GDP will decrease by 0.218%. GDP is defined as a function of underlying capital, labor and technology (generally unchanged), so this decrease reflects the combined effects of capital use (0.132%) and labor adjustments (−0.352%). There is a clear decrease in agricultural production in the Huaihe River basin, and the other three provinces will expand agricultural production and increase the inflow of agricultural commodities from outside at the same time. This boosts an increase in the real GDP of these three provinces, and the GDP of Henan, Jiangsu and Shandong separately increase by 0.109%, 0.147% and 0.112%. The increase of 0.109% in Henan is the combined influence of capital use (0.152%) and labor adjustments (0.042%). The increase of 0.147% in Jiangsu is the combined influence of capital use (0.179%) and labor adjustments (0.08%). The increase of 0.112% in Jiangsu is combined influence of capital use (0.148%) and labor adjustments (0.046%). GDP in 2020 will be 0.037% higher than the baseline scenario in the Huaihe River basin, which reflects the combined effects of capital use (0.403%) and labor adjustments (−0.396%). Agriculture is a labor-intensive sector, employment will be negatively affected in and the whole Huaihe River basin. A decrease in the amount of agricultural commodities will cause an increase in price. The consumer price index (CPI) will be 0.481% higher in 2020 than the baseline. As a result, household real consumption will be 0.174% lower and investment will be 0.296% higher than the baseline in 2020. The price of Anhui agricultural commodities rises relative to outside; the outflow of Anhui decreases by 0.41%, and the inflows increase. Changes in the inflow and outflow of the Huaihe River basin are respectively 0.023% and −0.119%.

Table 4 | Net VW flow matrix of the Huaihe River basin in 2020 (10^6 m^3)

	Henan	Anhui	Jiangsu	Shandong	Net VW outflows	Total VW outflows to other regions outside the basin	Total VW outflow
Henan	–	−148	614	797	1,263	3,071	4,334
Anhui	148	–	1,052	936	2,136	6,236	8,372
Jiangsu	−614	−1,052	–	1,043	−623	12,695	12,072
Shandong	−797	−936	−1,043	–	−2,776	4,652	1,876
Total	−1,263	−2,136	623	2,776	0	26,653	26,653

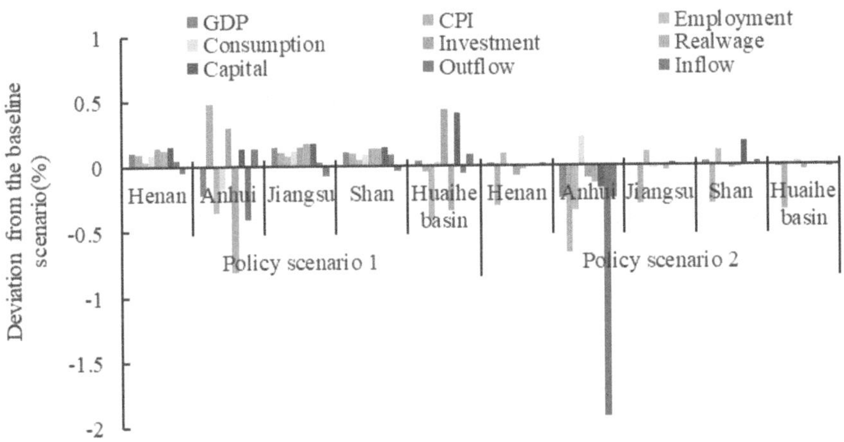

Figure 3 | Macroeconomic impact under two policy scenarios in 2020 (%). *Note:* % cumulative deviation from the baseline scenario. *Source:* simulation results.

In scenario S2, Anhui's agricultural outflow demand reduces by 5%, Anhui's GDP is negatively affected, falling by −0.248 in 2020. The decrease of outflow mainly comes from Anhui's agriculture, which is a labor-intensive sector, and employment will be negatively affected (−0.342%) in Anhui.

In S2, more agricultural production is available in Anhui markets, the price of agricultural produce decreases, and CPI shows a downward trend. This effect can affect other regions through inter-regional trade links, Jiangsu and Shandong need to inflow agricultural commodities from outside. As labor and capital move between regions, employment and capital will increase in Henan, Jiangsu and Shandong, leading to an increase in GDP in these three provinces. Employment and capital in Henan will increase 0.094% and 0.013%, lead to a 0.018% increase in GDP. For the Huaihe River basin, the positive influence of the Henan, Jiangsu and Shandong cannot offset the negative influence of Anhui, and the GDP of the Huaihe River basin shows a downward trend. But the impact range of GDP is smaller than that of Anhui.

Water consumption

In S1, the water consumption of Anhui decreases by 2.32% in 2020 (see Table 5), driving agricultural water consumption in the Huaihe River basin to drop by 0.65%. The agricultural output increases in Henan, Jiangsu and Shandong have a large demand for water, so water consumption is 0.08%, 0.10% and 0.69%, respectively, in 2020. Water consumption in most other industries shows a downward trend. Food processing and manufacturing is a downstream industry of agriculture, which results in a decrease in production due to the lack of agricultural products in local markets in Anhui.

In S2, agricultural water consumption reduces by 1.56%, resulting in agricultural water consumption reducing by 0.3% in the Huaihe River basin. Agriculture being the main water consumer, reductions in agricultural water consumption will relieve the water stress of other sectors, and the water consumption of other sectors will increase. Water consumption in food processing and manufacturing will increase by 1.96% due to its upstream industry output increasing. The agricultural output of Jiangsu and Shan will increase, so the water consumption of agriculture increase.

Virtual water flow

In S1, the VW of Anhui decreases by $322 \times 10^6\ m^3$ in 2020. The VW outflow to outside the basin decreases by $178 \times 10^6\ m^3$ (see Table 6). Because of Anhui decreasing outflows of agricultural commodities to Henan, Jiangsu and Shandong provinces, these three provinces need to inflow agricultural commodities from elsewhere. The VWT from Henan to Jiangsu and Shandong decreases by $25 \times 10^6\ m^3$ and $10 \times 10^6\ m^3$, respectively, and the net VW outflow of Henan decreases by $23 \times 10^6\ m^3$. The net VW outflow of Jiangsu and Shandong increases by $78 \times 10^6\ m^3$ and $89 \times 10^6\ m^3$, respectively, in the Huaihe River basin. But VW

Table 5 | The water consumption impact of different sectors under two policy scenarios in 2020 (%)

	S1 (%)					S2 (%)				
Sectors	**Henan**	**Anhui**	**Jiangsu**	**Shandong**	**Huaihe River basin**	**Henan**	**Anhui**	**Jiangsu**	**Shandong**	**Huaihe River basin**
Agriculture	0.69	−2.32	0.08	0.10	−0.65	0.00	−1.56	0.01	0.01	−0.30
Coal	−0.15	−0.17	−0.01	−0.01	−0.20	0.05	0.28	0.03	0.03	0.00
Petroleum	−0.21	−0.26	−0.01	−0.01	0.00	0.07	0.29	0.04	0.06	0.00
Other mining	−0.11	−0.25	−0.01	−0.01	0.00	0.09	0.32	0.10	0.09	0.00
Food	0.19	−0.33	0.03	0.03	−0.08	−0.22	1.96	−0.48	−0.23	0.70
Textile	−0.17	−0.40	−0.01	−0.01	0.00	−0.05	1.37	−0.09	−0.04	0.34
Papermaking	−0.3	−0.20	−0.01	−0.02	−0.26	0.04	0.69	−0.02	0.02	0.00
Chemistry	−0.31	−0.37	0.00	−0.01	0.00	0.01	0.48	0.01	0.02	0.12
Building materials	−0.11	−0.40	−0.01	−0.01	0.30	0.03	0.40	0.02	0.01	0.30
Metallurgy industry	−0.2	−0.29	0.00	0.00	0.00	0.05	0.36	0.08	0.10	0.13
Machine	−0.21	−0.23	0.00	0.00	0.00	0.09	0.35	0.04	0.09	0.00
Electronic industry	−0.38	−0.29	0.00	0.00	0.00	0.17	0.35	−0.01	0.03	0.00
Electricity	−0.22	−0.57	0.01	0.00	0.21	0.05	0.41	0.04	0.04	0.19
Other industry	−0.33	−0.09	−0.01	−0.01	0.00	0.05	0.62	−0.04	0.04	0.00
Construction	−0.13	−0.09	0.00	−0.01	0.00	0.00	0.12	−0.02	0.00	0.00
Transportation	−0.02	−0.02	0.00	−0.02	0.00	−0.01	−0.10	0.00	0.01	0.00
Wholesale-retail	−0.03	−0.02	0.00	−0.02	0.00	−0.01	−0.18	−0.01	0.00	0.00
Other services	−0.03	−0.04	0.00	0.00	0.00	0.01	0.05	0.00	0.01	0.00
Total	−0.35	−1.78	−0.07	−0.09	−0.14	0.00	−1.00	0.01	0.00	−0.21

Note: % cumulative deviation from baseline scenario. *Source:* simulation results.

Table 6 | The VW transfer of S1 in 2020 (10^6 m^3)

Outflows Inflows	Henan	Anhui	Jiangsu	Shandong	Net VW outflows	Total VW outflows to other regions outside of the basin	Total VW outflow
Henan	–	12	−25	−10	−23	−88	−111
Anhui	−12	–	−85	−47	−144	−178	−322
Jiangsu	25	85	–	−32	78	−106	−28
Shandong	10	47	32	–	89	−22	67
Huaihe River basin	88	178	−78	−89	0	−394	−394

Note: 10^6 m^3 cumulative deviations from the baseline scenario. *Source:* simulation results.

outflows to other regions outside of the basin decrease by 106×10^6 m^3 in Jiangsu and 22×10^6 m^3 in Shandong. The total VW outflow of the Huaihe River basin decreases by 394×10^6 m^3.

In S2, net VW outflows decrease by 89×10^6 m^3 in the basin, and the VW outflows to other regions outside of the basin decrease by 116×10^6 m^3 (see Table 7). Jiangsu and Shandong increase the inflows of agricultural commodities from Henan, and Anhui decreases agricultural commodities inflows from Henan. The VW transfer from Henan to Jiangsu and Shandong increases by 24×10^6 m^3 and 15×10^6 m^3, respectively. The VW transfer from Henan to Anhui decreases by 7×10^6 m^3. The net VW outflow of Henan decreases by 80×10^6 m^3. The net

Table 7 | The VW transfer of S2 in 2020 (10^6 m^3)

Outflows Inflows	Henan	Anhui	Jiangsu	Shandong	Net VW outflows	Total VW outflows to other regions outside of the basin	Total VW outflow
Henan	–	7	24	15	46	−80	−34
Anhui	−7	–	−46	−36	−89	−116	−205
Jiangsu	−24	46	–	0	32	−43	−11
Shandong	−15	36	0	–	21	−2	19
Huaihe River basin	−46	89	−32	−21	0	−241	−241

Note: 10^6 m^3 cumulative deviations from the baseline scenario. *Source:* simulation results.

VW outflow of Jiangsu and Shandong decreases by 43×10^6 m^3 and 2×10^6 m^3. The total VW outflow of the Huaihe River basin decreases by 241×10^6 m^3.

DISCUSSION

Water management and policy considerations

Unlike previous studies, this study uses a dynamic multi-region CGE model to provide an explanation of the macroeconomic, water consumption and VWT impacts on different regions and sectors. Two policy scenarios were set up, and S1 (Anhui's agricultural output reduced by 5%) can effectively reduce VW outflow and save more water resources. S1 causes an increase of 0.112% in GDP in the Huaihe River basin in 2020, but GDP in S2 decreases by 0.013%. However, employment declines in S1 and remains unchanged in S2. This finding calls for attention to the comprehensive impact (such as GDP, employment, consumption, food security) of inter-regional VW policies.

There is an interesting phenomenon: commodity trade and VWT flow from underdeveloped areas (Henan and Anhui) to relatively developed areas (Jiangsu and Shandong) and flow from areas with severe water scarcity to relatively water-rich areas. This phenomenon is a deviation from the concept of water-related sustainable development. The main causes are that many factors (such as coal and land) determine the distribution of economic production and productivity. The upstream areas of the Huaihe River basin have a large coal base and a major arable area, and these resources provide comparative advantages. In addition, the upstream areas occupy an absolute advantage in other mining, food, papermaking, building materials and metallurgy industry (these are water-intensive and pollution-intensive industries). Water trade is not only used as a water management strategy, but also uses a land or resource management strategy. VWT improves global land-use efficiency (Seekell 2011). This finding should call attention to the impact of inter-regional VW transfer on local water stress and ecological environment.

Impact of the South-North Water Transfer Project on VWT

The South-North Water Transfer Project (SNWTP) represents the largest-scale attempt to solve the uneven spatial-temporal distribution of water resources and relieve the North China Plain's water problems (Barnett *et al.* 2015). By March 20, 2019, the middle line of SNWTP had transferred a total of 20.5×10^9 m^3 water. Henan province received the largest amount of water: 6.942×10^9 m^3. Annual water transfer is increasing year by year. SNWTP alleviated the water shortage in Henan, but total VW outflows will reach 4.3×10^9 m^3 in 2020, which will further exacerbate water stress in Henan. The east line of SNWTP had continuously transferred a total of 3×10^9 m^3 water to Shandong, but total virtual water outflows will reach 1.8×10^9 m^3 in 2020, which will further exacerbate water stress in Shandong. SNWDP has a limited impact on alleviating water shortage, but it may exacerbate the VW outflow in water-scarce regions.

Limitations

There are several limitations that require future research. Firstly, only blue water was taken into consideration, and green and grey water (or water pollution) has been ignored. Secondly, 18 sectors were considered, but agriculture was not further split into crops, livestock and other agricultural products, forestry, and fishing, which could be corrected in the future by analyzing different agricultural sub-sectors. Thirdly, surface water and groundwater were not distinguished. It is important that the sectoral dependence on surface water and groundwater are distinguished in the future.

CONCLUSIONS

This study has progressed the following aspects. Firstly, the study is a meaningful attempt to analyze the role of VWT by using a dynamic multi-regional CGE. 18 sectors and four regions (Henan, Anhui, Jiangsu and Shandong) in the Huaihe River basin are included in the model. This allows the reader an in-depth appreciation of the impact of different VWT on the sectors and regions. The CGE is based on 12 years of data (2009–2020), which means long-term changes in spatial patterns of production, consumption, trade and VW flows can be studied. Secondly, the VW study is not only focused on agriculture, but also includes industry and service sectors. Thirdly, two hypothetical scenarios were set to analyze the economic, water consumption and VWT impact.

Our results show that the Huaihe River basin is a net VW outflow region, with a transfer of 26.653×10^9 m^3 VW in 2020; agriculture VWT accounts for 63% of the total VWT and therefore often decides the direction of VWT among regions. The industry sector is an important part of VWT, and affects the scale of VWT among the regions. The service sector's water use coefficient is small and does not have a great impact on VWT patterns. Jiangsu is the largest VW outflow province (main outflow is to outside), followed by Anhui. The VW outflows of these two provinces respectively are 12.072×10^9 m^3 and 8.372×10^9 m^3 in 2020. Anhui is a region with severe water stress, but is a net VW outflow province to Henan, Jiangsu and Shandong and outside of the Huaihe River basin. Henan and Anhui are net virtual water outflow areas, while Jiangsu and Shandong are net virtual water inflow areas in the Huaihe River basin. This reality contradicts the concept of sustainable water development. S1 (Anhui's agricultural output reduced by 5%) can effectively reduce VW outflow and save more water resources. S1 also causes an increase of 0.112% in the GDP of the Huaihe River basin in 2020, but S2 causes GDP to decrease by 0.013%. Employment declines in S1 but is unchanged in S2.

ACKNOWLEDGEMENTS

This work was funded by National Natural Science Foundation of China (52009042, 52025093), National Key R&D Program of China (2017YFC1502405, 2017YFC0405805-03) and State Key Laboratory of Simulation and Regulation of Water Cycle in River Basin (China Institute of Water Resources and Hydropower Research) open research funds (IWHR-SKL-KF201918).

DATA AVAILABILITY STATEMENT

All relevant data are included in the paper or its Supplementary Information.

REFERENCES

Allan, J. 1994 Overall perspectives on countries and regions. In: *Water in Arab World: Perspectives and Progress* (P. Rogers & P. Lydon, eds). Harvard University Press, Cambridge, MA.

Arto, I., Andreoni, V. & Rueda-Cantuche, J. M. 2016 Global use of water resources: a multiregional analysis of water use, water footprint and water trade balance. *Water Resour. Econ.* **15**, 1–14.

Barnett, J., Rogers, S., Webber, M., Finlayson, B. & Wang, M. 2015 Sustainability: transfer project cannot meet China's water needs. *Nature* **527**, 295–297.

Berrittella, M., Hoekstra, A. Y., Rehdanz, K., Roson, R. & Tol, R. 2007 The economic impact of restricted water supply: a computable general equilibrium analysis. *Water Resour.* **41** (8), 1799–1813.

Chapagain, A. K. & Hoekstra, A. Y. 2011 The blue, green and grey water footprint of rice from production and consumption perspectives. *Ecol. Econ.* **70** (4), 749–758.

Chen, Z. & Chen, G. 2013 Virtual water accounting for the globalized world economy: national water footprint and international virtual water trade. *Ecol. Indic.* **28** (5), 142–149.

Deng, G., Ma, Y. & Li, X. 2015 Regional water footprint evaluation and trend analysis of China – based on interregional input–output model. *J. Clean. Prod.* **112**, 4674–4682.

Dixon, P. & Rimmer, M. 2002 *Dynamic General Equilibrium Modeling for Forecasting and Policy: A Practical Guide and Documentation of MONASH*. North-Holland, Amsterdam.

Duarte, R., Pinilla, V. & Serrano, A. 2019 Long term drivers of global virtual water trade: a trade gravity approach for 1965–2010. *Ecol. Econ.* **156** (C), 318–326.

Hanasaki, N., Inuzuka, T., Kanae, S. & Oki, T. 2010 An estimation of global virtual water flow and sources of water withdrawal for major crops and livestock products using a global hydrological model. *J. Hydro.* **384** (3), 232–244.

Hoekstra, A. Y. 2003 Virtual water trade between nations: a global mechanism affecting regional water systems. *Glob. Change News Lett.* **54**, 2–4.

Hoekstra, A. Y. & Chapagain, A. K. 2008 *Chapter 8. The Water Footprint of Coffee and Tea Consumption: Globalization of Water: Sharing the Planet's Freshwater Resources*. Blackwell Publishing Ltd, Oxford.

Hoekstra, A. Y. & Hung, P. Q. 2005 Globalisation of water resources: international virtual water flows in relation to crop trade. *Glob. Environ. Change* **15**, 45–56.

Hoekstra, A. Y., Mekonnen, M. M., Chapagain, A. K., Mathews, R. E. & Richter, B. D. 2012 Global monthly water scarcity: blue water footprints versus blue water availability. *PLoS ONE* **7** (2), e32688.

Jiang, Y., Cai, W., Du, P., Pan, W. & Wang, C. 2015 Virtual water in interprovincial trade with implications for China's water policy. *J. Clean. Prod.* **87** (1), 655–665.

Kumar, M. D. 2018 Physical transfer of water versus virtual water trade: economic and policy considerations. *Water Econ. Policy* **4** (3), 1850001.1–1850001.21.

Lenzen, M. 2009 Understanding virtual water flows: a multiregion input–output case study of Victoria. *Water Resour. Res.* **45** (9), 318–326.

Mai, Y., Dixon, P. & Rimmer, M. 2010 *CHINAGEM: A Monash-Styled Dynamic CGE Model of China*. Centre of Policy Studies/Impact Centre Working Papers, pp. 28–30.

Roson, R. & Sartori, M. 2010 Water scarcity and virtual water trade in the Mediterranean. *SSRN Electron. J.* **28** (08), 875–899.

Robson, E. N., Wijayaratna, K. P. & Dixit, V. V. 2018 A review of computable general equilibrium models for transport and their applications in appraisal. *Transport Res. F.-Traf.* **116**, 31–53.

Seekell, D. A. 2011 Does the global trade of virtual water reduce inequality in freshwater resource allocation? *Soc. Natur. Resour.* **24** (11), 1205–1215.

Shi, C. & Zhan, J. 2015 An input–output table-based analysis on the virtual water by sectors with the five northwest provinces in China. *Phys. Chemi. Earth* **79–82**, 47–53.

Velázquez, E. 2007 Analyzing andalusian virtual water trade in an input–output framework. *Reg. Stud.* **41** (2), 185–196.

Wang, Z., Huang, K., Yang, S. & Yu, Y. 2013 An input–output approach to evaluate the water footprint and virtual water trade of Beijing, China. *J. Clean. Prod.* **42**, 172–179.

Wittwer, G. & Horridge, M. 2008 A multi-regional representation of China's agricultural sectors. *China Agr. Econ. Rev.* **1**, 420–434.

Zhang, W., Fan, X., Li, Y., Wang, S. & Chen, B. 2020 Spillover risk analysis of virtual water trade based on multi-regional input–output model – a case study. *J. Environ. Manage.* **275**, 111242.

Zheng, Y. & Fan, M. 2008 *China CGE Model and Policy Analysis*. Social Science Academic Press, Beijing.

First received 30 July 2020; accepted in revised form 9 December 2020. Available online 28 December 2020

Numerical simulation of three-dimensional hydrodynamic characteristics and pollutant diffusion behavior by ROMS model

Jing Ren, Chazhong Ge, Chunxu Hao, Rui Hu, Kefu Cui, Chen Peng, Quan Zhou, Aiyu Yu, He Xu, Shiqing Liu and Feng Li

ABSTRACT

Based on the ROMS (Regional Ocean Modeling System) model, a three-dimensional hydrodynamic model and a convection diffusion model are built in this paper. The three-dimensional tidal current model is intended to reflect the distribution characteristics of the tidal level and the flow field of water with different depths of the Bohai Bay. On this basis, a water exchange correlation matrix is adopted to analyze the water exchange characteristics of the offshore area of Binzhou City in the offshore planning functional area of Bohai Bay. Considering the functional planning of the offshore area of Binzhou city, the paper simulates how COD and NH_4^+-N in the water discharged from the sewage outfall of the city migrate and disperse under tidal current. In this way, the paper concludes the concentration distribution patterns of pollutants at different times. The results serve as basis and reference to delimitation of environmental management and control unit demarcation in Binzhou nearshore.

Key words | pollutant diffusion, ROMS, three-dimensional numerical simulation, water exchange characteristics

Jing Ren
School of Environment,
Liaoning University,
No.66, Chongshan Middle Road, Huanggu District, Shenyang,
China, 110036

Jing Ren
Chazhong Ge
Chunxu Hao (corresponding author)
Rui Hu
Chen Peng
Quan Zhou
Aiyu Yu
Chinese Academy of Environmental Planning of the Ministry of Ecology and Environment,
Beike Chuangye Building, No. 10, Dayangfang, Beiyuan Road, Chaoyang District, Beijing,
China, 100012
E-mail: *haocx@caep.org.cn*

Jing Ren
He Xu
School of Environment,
College of Environmental Science and Engineering, Nankai University,
No.38, Tongyan Road, Haihe Education Park, Jinnan District, Tianjin,
China, 300350

Kefu Cui
Central Research Institute of SIASUN Co., Ltd,
Xinsong Intelligent Industrial Park, No. 33, Quanyun Road, Hunnan New District, Shenyang,
China, 110168

Shiqing Liu
Binzhou Ecology and Environment Service Center,
No. 669, Bohai 18th Road, Binzhou,
China, 256600

Feng Li
Binzhou Ecology and Environment Monitoring Center,
No. 669, Bohai 18th Road, Binzhou,
China, 256600

HIGHLIGHTS

- Establish 3D hydrodynamic model of Bohai Bay.
- Establish water exchange correlation matrix to analyze nearshore water transport characteristics.
- Quantify the contribution of water exchange.
- Effects of river pollution load on nearshore water quality were simulated by our hydrodynamic model.
- Results serve as basis in delimitation of environmental management and control zone demarcation.

doi: 10.2166/ws.2020.380

INTRODUCTION

As China's coastal economy and marine industry gain traction in these years, the offshore area has been exposed to aggravating pollution due to rapid expansion of production activities, such as building ports, shipping transportation, building breakwaters, reclaiming land from the sea, protecting beaches, discharging pollutants into the sea, and harnessing estuaries and aquaculture. Pollutants can mix with surrounding waters through dilution, diffusion, convection and transportation after they go with the surface runoff into the ocean, which quickly brings down the concentration. Driven by the dual pressure from environmental protection and economic growth, research on the power system in the offshore area and the pollutant migration and diffusion pattern can make a great difference. Policies including Eco-functional Regionalization and Guidelines for Demarcation of Red Lines for Ecological Protection released by the Ministry of Ecology and Environment (hereinafter referred to as MEE) aim to guide localities through the principal function regionalization process, bringing a step closer the intensified protection of the ecological environment and natural resources in important regions and an optimized spatial development pattern (Li *et al.* 2019). After the 18th National Congress of the Communist Party of China, China has set out on a new mission of spatial management and control of sea areas to make progress in marine ecological red line and marine main function zoning across the board (Gao *et al.* 2018). What makes marine space management and control stand out lies in it being available to determine the structure and function of the ecosystem based on the physical, chemical and bio-ecological characteristics of the ecosystem in different locations, identify the important bio-ecological areas, and thus manage them in a differentiated way (Fock 2008). However, problems still exist such as relying on the original environmental function division and insufficient estimation of marine hydrodynamic diffusion conditions (Wang 2018) when delivering the work, especially for marine environmental control units division.

Three-dimensional hydrodynamic models become the mainstream of hydrodynamic numerical simulation as a better way to simulate what is really happening. Regarding water transport engineering and water environment application, the DHI Mike3 model, the Estuarine Coastal Ocean Model and the Princeton Ocean Model are all three-dimensional mathematical models based on static pressure assumption. In recent years, considering the realities of short wave driven flow, strong density gradient stratified flow and flow with large topographic changes, some achievements (Song *et al.* 1999; Koçyigi *et al.* 2002) have been made in the research into three-dimensional hydrodynamic mathematical models based on the non-hydrostatic assumption. Among them, ROMS (Regional Ocean Modeling System), as an open-source three-dimensional regional ocean model (Haidvogel *et al.* 2000), has been widely used in flow field simulation at different scales. It has obvious advantages in water level and flow field simulation caused by meteorological factors or astronomical tides, small-scale water physical movement simulation such as coastal ports, river channels, etc. (Budgell 2005).

Bohai Bay is a typical semi-closed bay. Its main hydrodynamic factors, wave and current (including tidal current, runoff, circulation, wave-generated nearshore current, etc.), interact with each other to shape a complex dynamic system. Its poor exchange capacity between the sea water and the outside world takes a toll on water quality of the offshore area as polluted water cannot spread rapidly. This paper takes the offshore area of Binzhou City in Shandong Province as the research object to analyze the dynamic factors in the ocean, adopting a hydrodynamic model based on the ROMS platform to simulate the water exchange characteristics of each area of the Bohai Bay and the migration and diffusion pattern of pollutants in the water. This can serve as a basis and reference for Binzhou city to demarcate environmental management and control units in offshore areas combined with how the marine environmental function zone and the marine ecological red line are identified.

METHODS

Construction of 3D hydrodynamic model

ROMS is an approximate solution to the N-S equation of Reynolds average under the Boussinesq assumption and the vertical static pressure assumption. Boussinesq assumes that differences in density within a body of water can be ignored unless the difference is caused by gravity. The vertical static pressure assumption holds that the depth scale in the coastal waters is much smaller than the horizontal scale, so the vertical equation of motion can be approximated by static equilibrium; that is, the vertical pressure gradient force and gravity balance. Cartesian coordinates are used horizontally and Sigma coordinates are used vertically.

The research takes the entire Bohai Sea area as the computational domain. To build an accurate three-dimensional hydrodynamic model of ocean scale, the open boundary of the calculation domain need to be accurate and stable. Besides, its continuous water level and velocity information should be easy to obtain. Thus, the open boundary of the tide on the east side is set between the National Marine Data Information Center at Tiger Beach (121°41′E,38°52′N) and Yantai Port (121°23′E,37°33′N). The data of moon phase, tidal hour and height of tide are obtained from the National Marine Data Information Center.

As Bohai Bay includes many ports in several provinces, data on human activities are difficult to estimate without reliable investigation; the effects of pollution caused by human were not quantified in this model.

The research adopts an orthogonal grid generated by the seagrid toolkit of Matlab. The total number of grid is 140 × 150, and the areas with large grid scales concentrate on the northern coastal area of Liaodong Bay, which has a relatively small driving effect on the tidal currents of Bohai Bay. This will not generate great effects on numerical simulation accuracy. Computational domain and computational grid are shown in Figure 1. Considering the overall shape characteristics of the Bohai Sea, the eastern open boundary extends from southwest to northeast; setting the grid along the open boundary is conducive to calculating the driving force. Besides, there are sparse grids in the northern part and a dense grid in the southern area, which can concentrate the calculation resources in the areas of our primary concern. At the beginning of calculation, water level and velocity of the model will adopt a zero initial condition. The water level of the opening boundary will adopt the harmonic constant interpolation of eight tidal constituents from Tiger Beach to Yantai, including M2,S2,K1,O1,N2,P1,K2 and Q1. Based on the wind rose chart of Binzhou, the model simulates the characteristics of the monsoon, continuously changes the wind direction, then calculates the average wind speed of each specific direction as a wind speed of 10 m above the sea surface, to estimate the influence of wind on water flow. After verification, it shows that the error of harmonic water level is very small. Hence, we can use the harmonic water level to replace the tidal table water level as the tidal boundary of the hydrodynamic model. The major parameters of the model are set as follows:

(1) Time step. Limited by CFL conditions, the time step of the internal model of the tide numerical model is 300s, and the time step of the external model is 15s. The water level information of the east side opening boundary will be read every time step.
(2) Bottom calculation parameter. Since the sediments of Bohai Bay are mainly silt and clay, the granularity is relatively small, the roughness length is 0.005 m, and the drag coefficient of bottom friction is 0.0008.
(3) Set up appropriate vertical calculation parameters, which mainly include a vertical mixing coefficient, momentum vertical mixing coefficient, surface tensile parameter and bottom tensile parameter.

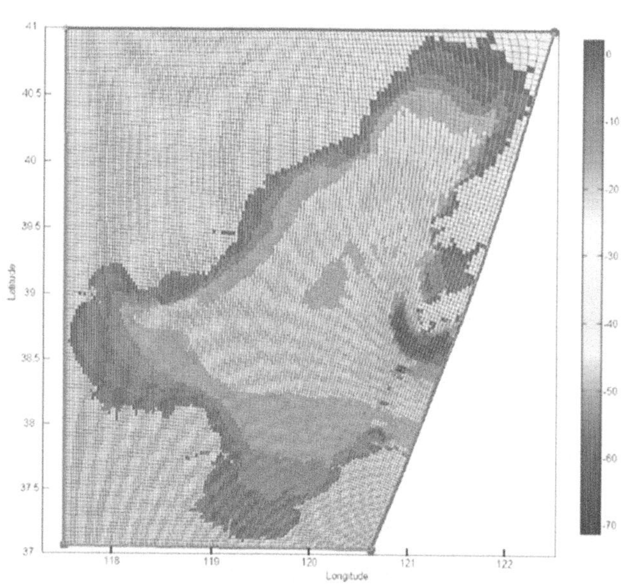

Figure 1 | Water depth distribution of research area and orthogonal grid setup.

RESULTS AND DISCUSSION

The status quo of environmental quality in Binzhou offshore areas

The northern coast of Binzhou City, Shandong Province, is located in the South Bank of Bohai Bay, with gentle slope and a shallow water, wide span of intertidal zone. Southwest wind is the predominant wind direction, featuring continental climate. Many rivers, such as Dakou River, Majia River and Taoer River, cross into the sea. The tidal current in the Binzhou waters are the regular semidiurnal tides. The rising tide flows to the southwest and the falling tide flows to the northeast. The reversing current is the typical tidal movement (Liu 2020). The Binzhou sea area is open to the east and north, with large waves, mainly wind and waves, supplemented by swell. Binzhou is home to a vast sea area, rich coastline resources, broad continental shelf and endowed natural environment.

At present, the Binzhou nearshore area is mainly developed for the sake of mariculture, raw salt production, port shipping, shipping industry, electric power industry, oil and gas exploitation, urban construction, road and bridge and coastal protection (Tian 2018). The water quality of Binzhou nearshore in 2017 and 2018 is shown in Table 1, dominated by the heavily and moderately polluted sea areas respectively. The results of the seawater quality monitoring data in the past five years show that inorganic nitrogen, total nitrogen, total phosphorus and other nutrient pollutants exceed the limits, and the eutrophication phenomenon is most visible around March every year in offshore areas of Binzhou.

Flow field feature analysis of Bohai Bay

It can be seen from different levels of velocity that in jet stream waters, the vertical differences of velocity are relatively evident. However, in areas with relatively small velocity, the velocities of surface, middle and bottom have almost no differences (see Figure 2). Besides, the simulation flow result shows the influences of nearshore engineering and natural structures on the direction of current motion (Zhou & Yang 2019; Li *et al.* 2020). In general, the flow field of Bohai Bay is characterized by a regular east-west reciprocating flow. In the northern part, due to the blocking of the coast line of Caofeidian, Tangshan (Point A), the tidal currents wind along the coast line. The northwestern area

Table 1 | Areas of sea water with different levels of quality in Binzhou (2017–2018)

	Sea area (km^2)				
Sea water quality	**I**	**II**	**III**	**IV**	**Inferior to IV**
2017 average	130.22	458.53	465.19	318.77	627.29
2018 average	390.86	285.89	616.51	275.07	431.66

takes Nanjiang Port (Point B) of Tianjin as its boundary. When the tide rises, the flow direction of waters in northern Nanjiang Port is north by northwest, while the flow direction of waters in southern Nanjiang Port is mainly west, and the flow direction of some areas is slightly south. When the tide falls, the flow direction is the opposite.

By comparing flow directions of different depths at the same time, we can see that the flow directions of the surface, middle and bottom of Bohai Bay are basically the same. From the velocity at high tide, the velocity of northern Bohai Bay is greater than that of the southern area. On the east side of Caofeidian, the flow velocity of the surface, middle and bottom level is decreasing, while on the west side of Caofeidian, where streaming is generated, the velocity of tidal currents increases dramatically. The tidal currents then flow to the inshore region of Tianjin Binhai New Area, with a decreasing velocity. There is little difference in velocity in most of central area of Bohai Bay; at the inshore region of east Huangye Port (Point C) of Cangzhou, the velocity of tidal currents gradually decreases. At low tide, the velocity in most areas of Bohai Bay is similar to that of high tide. However, the velocity feature of streaming near Caofeidian is opposite to that of high tide, and the greatest velocity appears at the east side of Caofeidian.

Water exchange feature of Bohai Bay

In general, the water exchange capacity of Bohai Bay is rather poor. The southeast region has the largest semi-exchange period. The water exchange capacity of the northwest and north beach area is relatively poor, which is adverse to pollutant discharge. In the wave breaking region, pollutant transport and diffusion is characterized by parallel shoreline diffusion, so the pollutants cannot diffuse to open seas along with tidal currents. The pollutants emitted outside the wave-breaking region may generate influences on the offshore area in the south part of the emission point.

Based on water exchange relational matrix analysis (Sun 2007) and the coastline of Bohai Bay, combined with characteristics of nearshore human activities, we divided Bohai Bay into six subdomains (Figure 3). District I represents the southwest part of Bohai Bay, including the offshore sea of Binzhou. District II is the southeast part of the computational domain, namely the east part of Binzhou. District III and IV represent

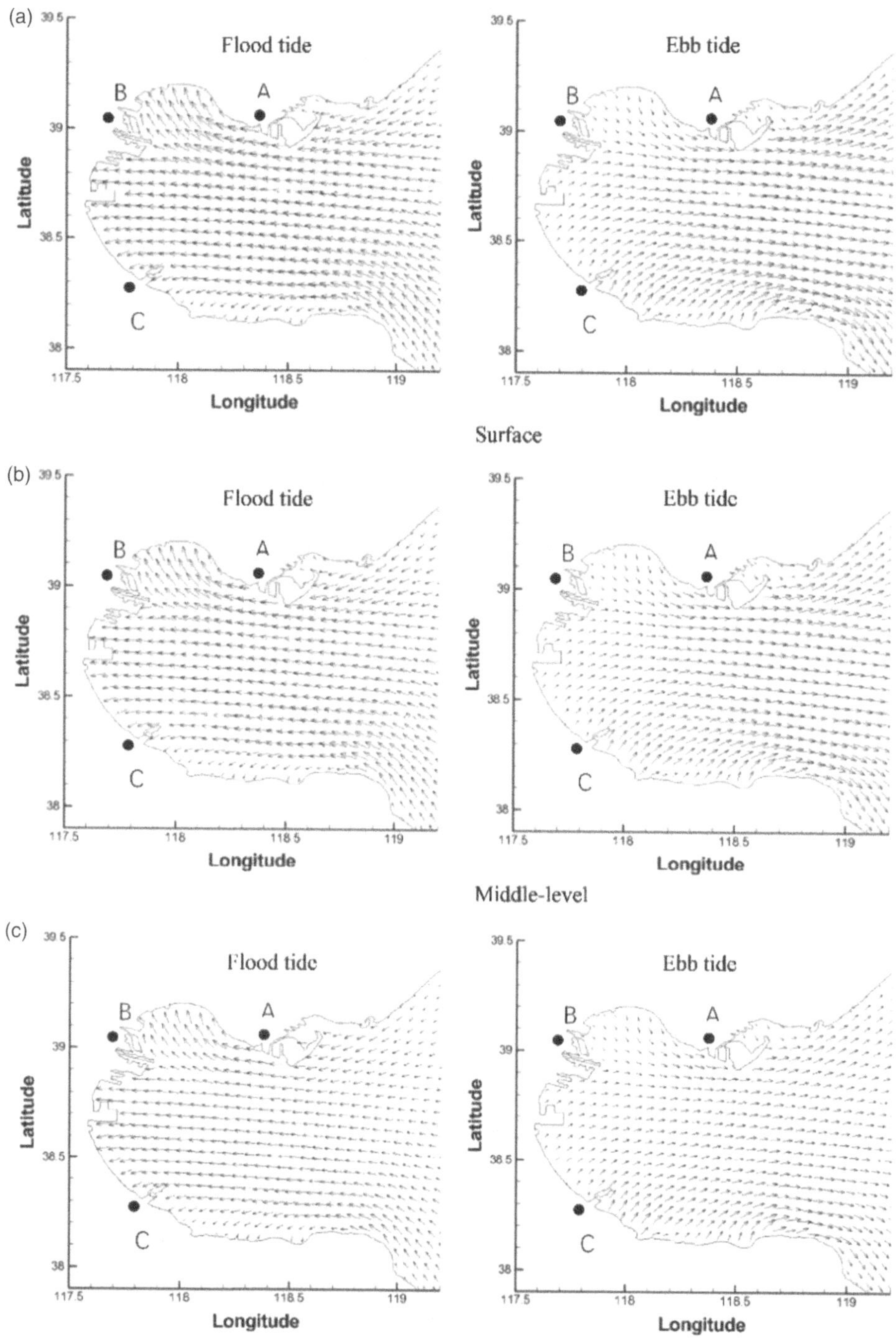

Figure 2 | Vector diagram of velocity at high tide and low tide in Bohai Bay. *Note*: Point A is Caofeidian of Tangshan, Point B is Nanjiang Port of Tianjin, Point C is Huangye Port of Cangzhou.

the middle west and middle east part of the computational domain respectively. District V and VI are the northwest and northeast part of the computational domain respectively. We simulated the water-exchange features of six subdomains, and the simulation period is a month. The simulation result is shown in Figure 4.

Figure 4(a) indicates that the original water of District I continuously flows out via water exchange to other

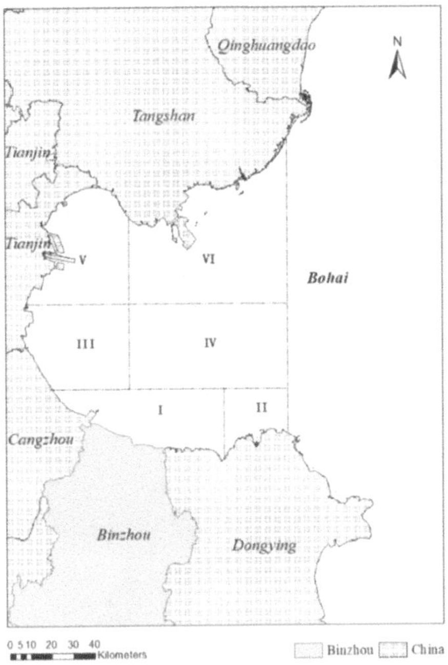

Figure 3 | Bohai Bay subdomain division plan.

subdomains. Six days after the simulation, the water exchange rate of this region exceeds 50% for the first time. On the 30th day, the original water of this region only accounts for 13.7% of the current water. During flow field movement, the original water of this region mainly flows to District III and IV, namely the middle west and middle east part of the computational domain; the water in this region contributes little to the output to District II, V and VI during water exchange. Only less than 1% of the water in District I flows to District VI. Therefore, during the simulation period, the sea water quality of Binzhou offshore sea has little influence on the offshore sea of Dongying, Shandong Province (District II), and the offshore sea of Tianjin (District V). And it has almost no influence on the offshore sea of Tangshan (District VI).

Figure 4(b)–4(e) show the proportion of the water exported to District I after water exchange in District II ~ VI to the original water in each district. It can reflect the influences of sea water quality of each region on the water quality of Binzhou offshore sea. According to the simulation result, the original water of the offshore sea of Dongying, Shandong Province, flows out to Binzhou at a relatively large amount; the sea water quality of Tianjin offshore sea and Tangshan offshore sea has little effect on the water quality of Binzhou. The offshore sea of Tangshan, Hebei Province almost has no influence on the water quality of Binzhou waters. It is worth noting that the results of the water exchange feature can explain the offshore seawater exchange pattern between Binzhou and surrounding cities. The water exchange feature should not be completely equivalent to the impact of pollution caused by surrounding cities.

Simulation of pollutant diffusion in offshore sea of Binzhou

The offshore sea of Binzhou includes Wuli County and Zhanhua District. Wuli County has three seagoing rivers, which are Dehuixin River, Majia River, and Zhangweixin River. The coordinates of the sea access monitoring section are E117.6871°/N37.9914°, E117.5987°/N37.9731° and E117.5641°/N38.0502° respectively; there are two sea access drain outlets in Zhanhua District, the coordinates of which are E 118.2353°/N 37.8044° and E 118.2163°/N 37.9970°.

Based on the above grid, we established a simulation environment of Binzhou offshore sea, wind velocity was set up according to the average wind direction and speed provided by Binzhou Meteorological Bureau. Based on the background values of COD and the NH_4^+-N indicator of Binzhou offshore sea from 2017 to 2018 (Ecological Environment Department of Shandong Province 2019), we simulated the influences of pollution discharge from the offshore sea access drain outlet and upstream rivers on the water quality of Binzhou offshore sea within 30 days, so as to investigate the potential for improving the marine environment. Since there is no reliable data to support and train the model, the transport/absorption/decomposition of pollutants by marine animals and plants are not taken into account in this paper. The model uses inert tracer particles to simulate pollutants, which can simulate the migration and diffusion of pollutants under physical conditions.

The result shows that the diffusivity of pollutants along the coastline is poor. This is mainly because the sea is making a reciprocating motion with the tide, leading to accumulated pollutants along the coast line (Figure 5). According to the marine environment function planning of Binzhou, this region is close to the harbor approach. Therefore, production activities have great effects on the water quality in this region. Besides, this region is close to the

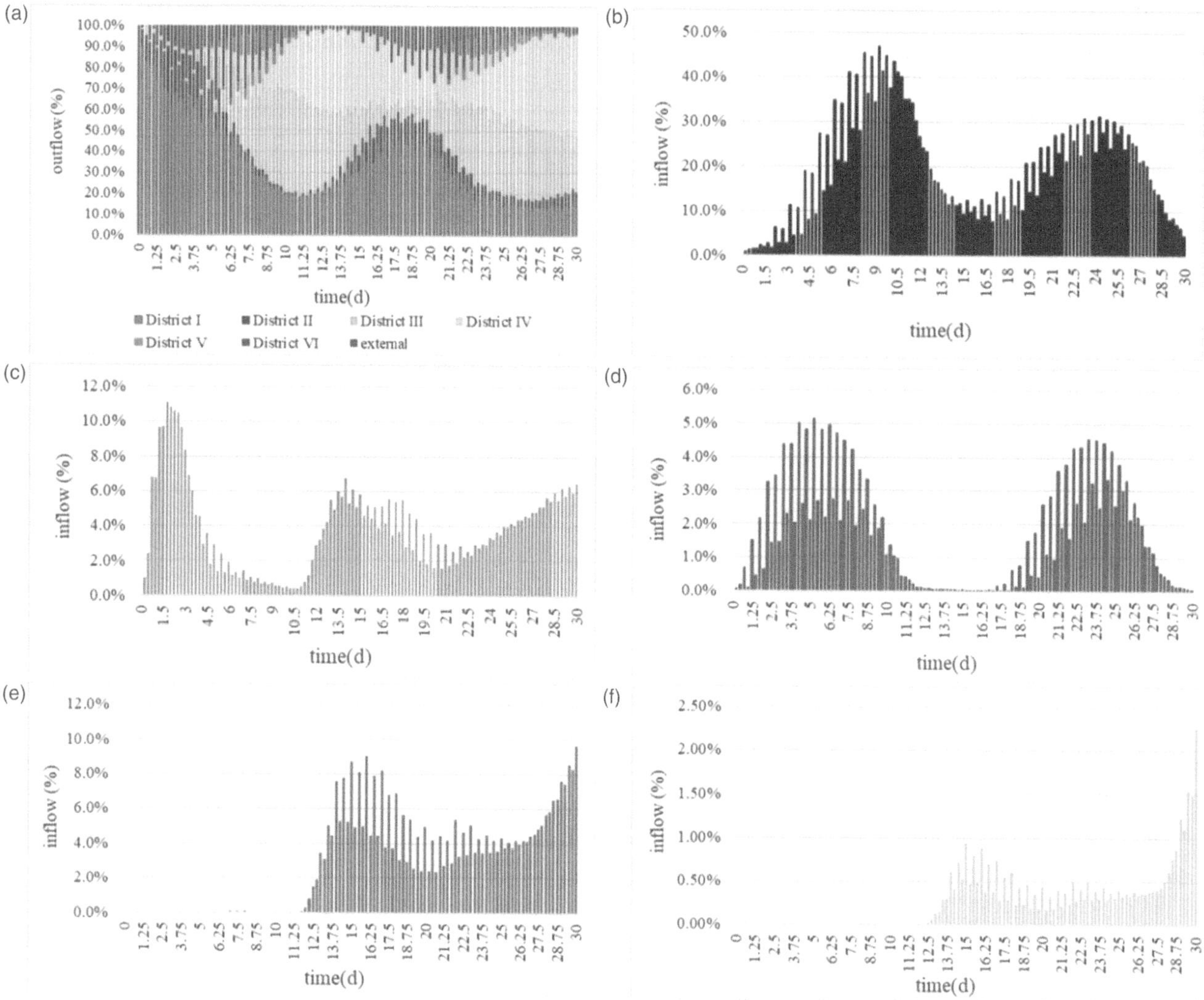

Figure 4 | Water exchange simulation result of District I and other subdomains. *Note*: (a) represents water exchange outflow result of District I; (b), (c), (d), (e) and (f) represent water exchange inflow results from District II, III, IV, V and VI to District I.

offshore sea, so at the entrance drain outlet, the emission concentration of COD in discharged water (from Wuli County, Zhanhua District Industry Park and factories) has a significant effect on the pollutant concentration of the offshore sea. If no water is discharged from the entrance drain outlet of Binzhou, then within 30 days of simulation, the COD concentration in the offshore sea can decrease by 19.11% (Figure 6). As for the NH_4^+-N, the potential reduction excent is 10.85%. According to the water quality monitoring results of Binzhou offshore sea in 2017 and 2018, the simulated pollutant concentration accumulation area agrees well with the water quality monitoring results of Binzhou offshore sea.

CONCLUSIONS

(1) *Publish technical guidelines of offshore Marine Control Unit division*

This research used ROMS to construct the hydrodynamic model, to simulate the water exchange features of Bohai Bay. In the simulation, we considered the influences of wind speed, tide level and other factors, preliminarily determined the area with poor diffusivity in Binzhou offshore sea, and used it as the basis for the Binzhou marine control unit division. In China, the division of Environmental Control Areas is in progress; meanwhile, no unified division method for

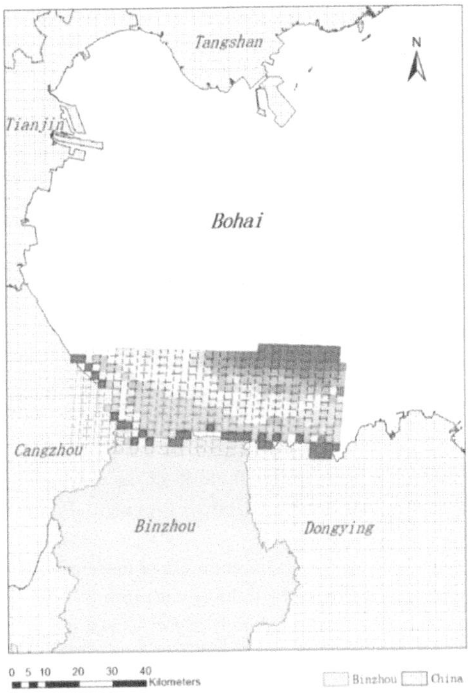

Figure 5 | Simulation of pollutant diffusion in Binzhou offshore sea.

Marine Control Areas has been published yet. In the demarcation of Key Marine Control Areas, it is necessary to identify weak dispersion regions in the nearshore. It has been verified by our study that ROMS has considerable advantages in this respect. We shall carry out in-depth research on pollutant diffusion behavior patterns (Chen *et al.* 2017.), combined with ocean vertical exchange analysis, and establish a better offshore sea diffusion model, so as to provide a theoretical basis for marine control unit division.

(2) *Accelerate the implementation of sea space control division results, improve supporting control system*

Binzhou should accelerate the ratification of marine control unit and marine ecology red line division results, implement marine ecology red line control requirements, accelerate the construction of a supporting control system (Huang *et al.* 2016), improve water quality monitoring capacity for the offshore sea, improve the marine control unit performance appraisal system, carry out research on an ecological compensation system for the red line region, so as to guarantee major progress in ecological environmental protection and resources conservation in the offshore sea.

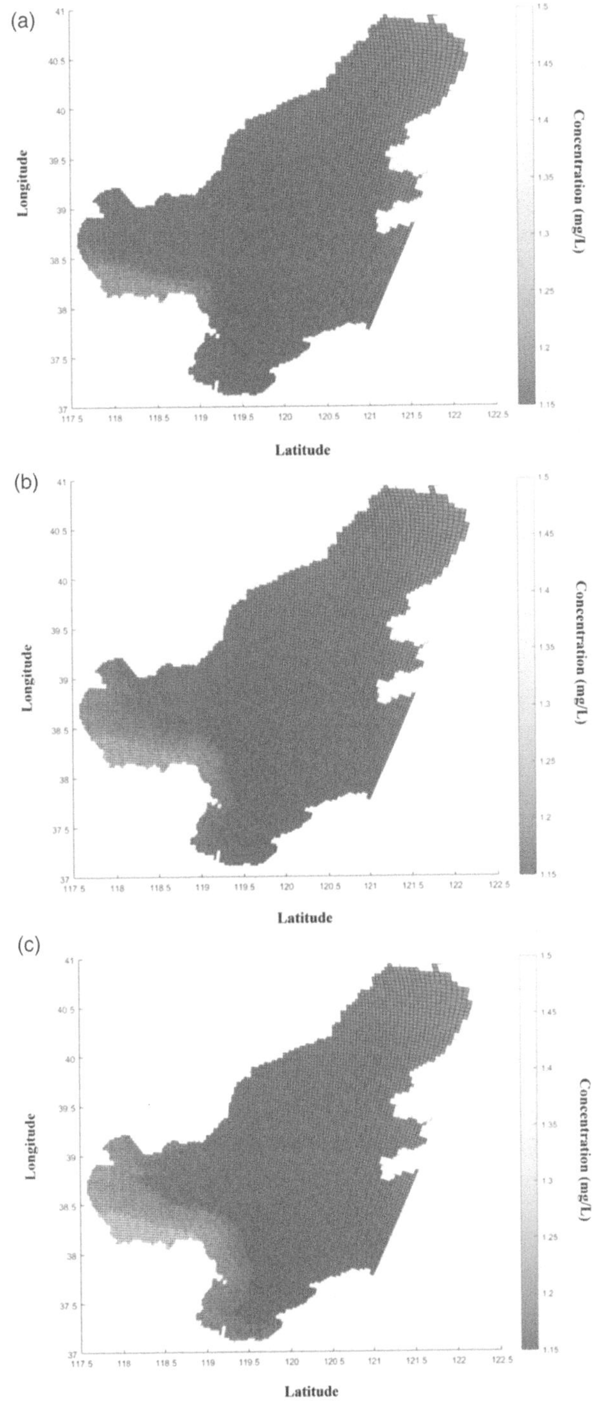

Figure 6 | Simulation of pollutant reduction potential in Binzhou offshore sea. *Note*: (a)-(c) represent COD concentration in the offshore sea in 5 days, 15 days and 30 days, respectively.

(3) *Strengthen coast line management, strictly control land reclamation projects*

The simulation results of water exchange features

shows that the water exchange capacity of Bohai Bay is poor in general, especially the southeast area, which has the largest semi-exchange period and is not conducive to pollutant discharge. Ocean development processes such as getting soil from the coast, hydraulic reclamation and burying will generate negative influences on hydrodynamic conditions of the offshore sea, increasing the risks of typhoons, storm tides and other marine disasters (Sun *et al.* 2020). We should accelerate top level design of offshore sea development control, strictly control land reclamation and introduce marine emerging industry projects, to scientifically and rationally promote the sustainable development of the offshore sea economy.

ACKNOWLEDGEMENTS

The authors would like to thank Binzhou Ecology and Environment Service Bureau for help on management recommendations. The work was supported by funding from the National Financial Program 'the Red Line of ecological protection, the bottom line of environmental quality, the upper line of resource utilization and the negative list of environmental admittance' Pilot Construction and Implementation of Ecological and Environmental Protection Project (2110107).

DATA AVAILABILITY STATEMENT

All relevant data are included in the paper or its Supplementary Information.

REFERENCES

Budgell, W. P. 2005 Numerical simulation of ice-ocean variability in the Barents Sea region. *Ocean Dynamics* **55** (3), 370–387.

Chen, G., Hu, W., Chen, B., Ma, Z. & Liao, J. 2017 Marine spatial planning using in marine ecological red line zoning – a case study of Dongshan County. *Journal of Applied Oceanography* **36** (01), 6–15.

Ecological Environment Department of Shandong Province 2019 *Environmental Status Bulletin of Shandong Province.* Available from: http://zlc.sdein.gov.cn/zkgb/ (accessed 5 June 2019)

Fock, H. O. 2008 Fisheries in the context of marine spatial planning: defining principal areas for fisheries in the German EEZ. *Marine Policy* **32** (4), 728–739.

Gao, Y., Zeng, J., Huang, W., Chen, Q., Shou, L. & Xu, X. 2018 The relationship between marine functional zoning and marine ecological red line. *Ocean Development and Management* **35** (01), 33–39.

Haidvogel, D. B., Arango, H. G., Hedstrom, K., Beckmann, A., Malanotte-Rizzoli, P. & Shchepetkin, A. F. 2000 Model evaluation experiments in the North Atlantic Basin: simulations in nonlinear terrain-following coordinates. *Dynamics of Atmospheres and Oceans* **32** (3), 239–281.

Huang, W., Zeng, J., Chen, Q., Du, P., Tang, Y. & Yang, H. 2016 Preliminary research on the zoning method of the marine ecological red line: a case study of Hainan Province. *Acta Ecologica Sinica* **36** (01), 268–276.

Koçyigi, M. B., Falconer, R. A. & Lin, B. 2002 Three-dimensional numerical modelling of free surface flows with non-hydrostatic pressure. *International Journal for Numerical Methods in Fluids* **40** (9), 1145–1162.

Li, Q., Wang, Z., Liu, Y., Li, W., Zeng, S., Xiang, W. & Xu, K. 2019 Research on Beijing-Tianjin-Hebei region environment control division and different access. *Environmental Impact Assessment* **41** (01), 28–33.

Li, Z., Liu, A. & Mo, X. 2020 A numerical method for solving 3D marine dynamics model based on surface state parameter. *Journal of Tropical Oceanography* **39** (01), 12–19.

Liu, S. 2020 Spatial distribution characteristics of heavy metals in soil and bottom sediments in Binzhou Coast Zone. *Shandong Land and Resources* **36** (01), 7–12.

Song, Z., Xue, H., Yan, Y., Mao, L. & Xu, F. 1999 Quasi-3D numerical simulation of tidal hydrodynamic field. *China Ocean Engineering:English Version* **3**, 265–276.

Sun, J. 2007 *The Relation Matrix Method and Its Application of Water Exchange Study for Sea Bay and Coastal Water.* Tianjin Univesity, Tianjin, China.

Sun, Y., Qiao, Y., Liu, C., Yu, H. & Wang, W. 2020 Marine ecological pressure and countermeasure in Tianjin. *Ocean Development and Management* **37** (02), 58–62.

Tian, G. 2018 Research on marine ecological environment quality anaysis and comprehensive management – Binzhou. In *Ecological Economy and Healthy China-China Ecological Economics Society 2018 Academic Annual Conference*, Xi'an, Shanxi, China.

Wang, S. 2018 Optimization of evaluation method of marine resources and environmental carrying capacity based on scientific management and control. *Marine Environmental Science* **37** (04), 608–612.

Zhou, Z. & Yang, N. 2019 Study on dynamic simulation of river water quality based on three-dimensional numerical model. *Environmental Protection Science* **45** (06), 108–113.

First received 18 August 2020; accepted in revised form 15 December 2020. Available online 30 December 2020

An analysis on the influence of precipitation infiltration on groundwater under different irrigation conditions in the semi-arid area

Long Yan, Min-jian Chen, Peng Hu, Di-li Li and Yong Wang

ABSTRACT

The West Liaohe Plain is a typical semi-arid area, where the process of rainfall infiltration to replenish groundwater is a key link in its vertical hydrological cycle. In this paper, we compare and analyze the impact upon soil moisture movement and water infiltration after the shift of irrigation method from flood irrigation to mulched drip irrigation under mulch through setting up in-field in-situ observation points and carrying out groundwater depth dynamic observation. The results show that compared with mulched drip irrigation under mulch, flood irrigation has a stronger response to rainfall infiltration and a quicker response time in the rise of underground depth. With the decrease of groundwater level, the effect of rainfall infiltration to replenish groundwater is significantly weakened. In the flood irrigation area, the groundwater depth at about 8 m already has no obvious response to a small amount of rainfall. However, the groundwater depth at 6 m in the area of mulched drip irrigation under mulch already has no response to rainfall. Therefore, when groundwater extraction is carried out in irrigation areas, reasonable groundwater extraction levels should be designated in light of different irrigation methods to maintain the sustainable utilization of groundwater.

Key words | groundwater level, irrigation conditions, precipitation infiltration, soil moisture content

Long Yan
Min-jian Chen
Peng Hu (corresponding author)
Di-li Li
Yong Wang
China Institute of Water Resources and Hydropower Research,
Beijing 100038,
China
E-mail: *hp5426@126.com*

Long Yan
State Key Laboratory of Simulation and Regulation of Water Cycle in River Basin,
Beijing 100038,
China

HIGHLIGHTS

- Observed rainfall infiltration realized through flood irrigation and mulched drip irrigation under mulch.
- Analyze the mechanism of the influence of irrigation conditions on rainfall infiltration.
- Analyze the change of soil moisture content.
- Analyze their impacts upon rainfall infiltration.
- The promotion of mulched drip irrigation under mulch will exert a significant impact on rainfall infiltration.

INTRODUCTION

The impact of human activities and of climate change is exacerbating the contradiction between the supply and demand of water resources in various regions (Xia & Shi 2016). As the contradiction becomes more and more prominent, human beings continue to develop towards high-efficiency water-saving agriculture. At the same time, when water-saving measures are contributing to the increase of crop yield, we should also pay attention to the impact that these water-saving measures will exert on groundwater replenishment in related regions. For instance, with the

doi: 10.2166/ws.2021.011

popularization of refined irrigation measures such as sprinkler irrigation and micro-drip irrigation, measures like concrete lining adopted in various levels of the channel system in the water delivery and distribution system have played a positive role in reducing deep water leakage and water loss along the channels and have achieved the purpose of improving the utilization rate of irrigation water in irrigation areas (Arbat *et al.* 2008; Bufon *et al.* 2012; Skaggs *et al.* 2012; Qin *et al.* 2016). Meanwhile, however, water replenished to groundwater in related regions from irrigation return water has also decreased accordingly. In addition, the popularization of high-yield and high-quality crop varieties as well as the implementation of high-efficiency fertilization technologies have led to a significant increase in the output of biomass per unit area, which has also led to a substantial consumption of soil water. This means substantial decrease in the amount of underground water replenished from rainfall and irrigation water. In their researches on water quantitative transformation relationship between crop structure, crop yield, irrigation system, etc. and water quantity, Xuzhou Hanwang Hydrological Experiment Station and Baoding Ranzhuang Water Resources Experiment Station pointed out that high-efficiency agricultural water utilization measures improve hydrological utilization efficiency along with their negative impact on groundwater replenishment (Wei & Ba 2003).

Traditionally, flood irrigation has always been adopted for crops in the West Liaohe Plain. Since 2005, mulched drip irrigation under mulch has been promoted and applied as an efficient water-saving irrigation facility in the area around Tongliao City. A stable area of approximately 40,000 mu of farmland was applied with mulched drip irrigation under mulch from 2008 to 2011, accounting for 0.43% of the total effective irrigation area. With the advancement of the Water Saving and Yield Increase Action Plan, the corn farmland area with mulched drip irrigation under mulch had dramatically increased from 344,000 mu in 2012 to 1.678 million mu in 2015, accounting for 17.1% of the total effective irrigation area. As of 2015, a total of 1.311 million mu of farmland with mulched drip irrigation under mulch had been established in Horqin District, Horqin Zuoyi Middle Banner and Kailu County, accounting for 78% of the total farmland area with mulched drip irrigation under mulch across the whole city. The whole region has become a major area where mulched drip irrigation under mulch is mainly applied.

Rainfall is an important replenishment factor for vertical hydrological cycle in semi-arid areas (Yan *et al.* 2018). Aiming at the problem of groundwater recharge by precipitation infiltration, many scholars have studied the spatial variability and influencing factors of precipitation infiltration coefficient (Shi *et al.* 2007; Huo & Jin 2015). Zhang *et al.* (2007) analysed the influence of unsaturated zone thickness on precipitation infiltration for recharge of groundwater; the result shows that following the continuous dropping of the groundwater table, the recharged rate tends to decrease in a limited period due to the increase in the infiltration path length. Rainfall infiltration replenishment for underground water is a key link in the process of the vertical hydrological cycle, but changes in irrigation conditions in farming irrigation areas will have an impact on this key link. Therefore, it is necessary to observe rainfall infiltration realized through flood irrigation and mulched drip irrigation under mulch respectively and analyze their impacts upon rainfall infiltration because of the change of irrigation method.

MATERIALS AND METHODS

Study areas

The West Liaohe Plain is located in the eastern part of the Xiliao River Basin, with its major part in Tongliao City, and lies between 41°05′ to 45°13′N and 116°10′ to 123°35′E. Its area is approximately 65,000 km^2. Within this region, the annual average precipitation is less than 380 mm, and the annual average water surface evaporation is more than 1,000 mm. Typically characterized with semi-arid features, this area normally has no direct surface runoff after rainfall. Vertical precipitation infiltration is an obvious phenomenon here, the groundwater is abundant and vertical recharge for groundwater is stable. As a relatively intact watershed within the farming-pastoral zone in the semi-arid area, the West Liao River has hydrological cycle characteristics that typically belong to the farming-pastoral zone in the semi-arid area. Its plain area fully exhibits the rainfall-evaporation-infiltration cycle, especially the process of rainfall infiltration to replenish groundwater. In recent years, the irrigation mode in the West Liaohe Plain has shifted from flood irrigation to mulched drip irrigation under mulch. Therefore, this paper takes the West Liaohe Plain as an example to study the impact of changes in irrigation method upon groundwater in a semi-arid farming-pastoral zone (Figure 1).

With the continuous expansion of the irrigated area, the increase in the number of irrigation wells, and the over-exploitation of groundwater, the groundwater depth shows a declining trend. This paper collected and sorted out the groundwater depth data of 74 groundwater monitoring wells in West Liaohe River Plain since 1980. In 1980, the average groundwater depth of West Liaohe River Plain was 2.33 m, and dropped to 7.55 m by the end of 2018, with an annual decline rate of 0.13 m/a (Figure 2). Before 2000, the downward trend of groundwater depth was not obvious, and the groundwater depth was between 2 and 4 m. After 2000, the groundwater depth decreased significantly, and the average groundwater depth in the plain area decreased by about 5 m. The most obvious decrease of groundwater burial depth was shown in Horqin District

Figure 1 | Study area. The map were created using ArcGIS 10.2 (http://www.esri.com/sofware/arcgis/arcgis-for-desktop).

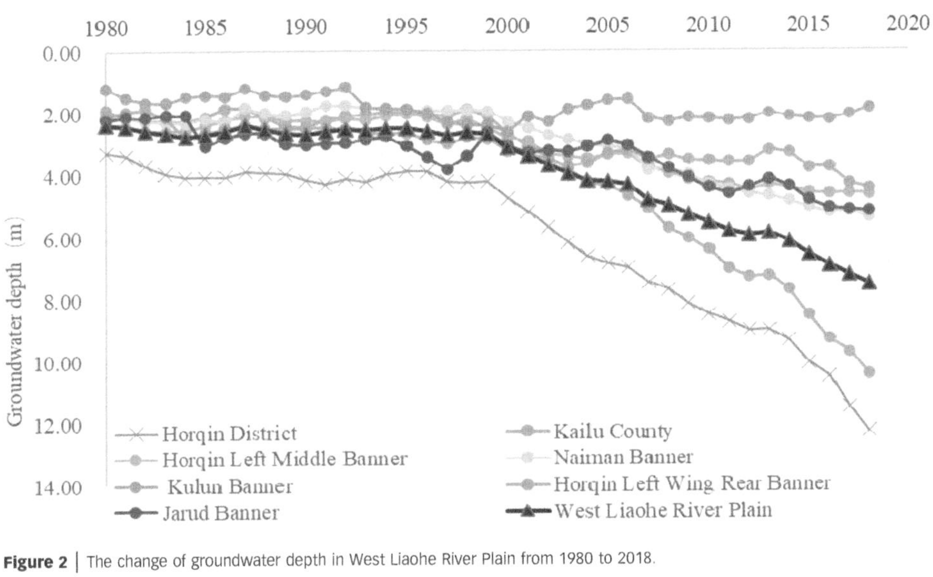

Figure 2 | The change of groundwater depth in West Liaohe River Plain from 1980 to 2018.

and Kailu County, two counties with the largest distribution of irrigated areas. The groundwater depth in Horqin area showed the maximum changes, falling from 3.2 m in 1980 to 12.29 m in 2018.

Mechanism analysis

Rainfall in the West Liaohe Plain basically does not lead to surface runoff, so replenishment for groundwater mainly comes from rainfall. This area has the rainfall-evaporation-infiltration vertical hydrological cycle characteristics. The farming irrigation area in the plain area is the guarantee for the social and economic development of the West Liaohe Plain. According to the development of agricultural and grazing industry in the West Liaohe Plain, the artificial ecological environment marked by the farming irrigation area has played a key role in this area. However, within the farming irrigation area of the West Liaohe Plain, rainfall cannot meet the water demand for the growth of crops, and irrigation has become a necessary means for the development of agriculture. Traditional flood irrigation has always been the main irrigation method in this area, and the return water produced in this way has played a certain role in the replenishment of groundwater in the area.

In terms of replenishment, atmospheric precipitation is intercepted by vegetation and falls to the ground surface to form net rainfall. Net rainfall infiltrates into the soil to replenish it with water. Part of the water can be transported through the unsaturated zone to the groundwater surface and replenish with underground water (Rimon *et al.* 2007). Under the condition of flood irrigation, irrigation return water will be produced and re-infiltrate to replenish with underground water. In terms of discharge, the whole process consists of interception evaporation, water surface evaporation, soil and vegetation evapotranspiration, and artificial extraction, among which vegetation transpiration and soil evaporation are reflected in unsaturated zone evapotranspiration and phreatic evaporation. As groundwater enjoys an absolute advantage in the composition of water resources in the West Liaohe Plain, and agricultural development must rely on the extraction of groundwater for irrigation, artificial extraction is the largest groundwater discharge item in the West Liaohe Plain. To maintain the stability and renewability of groundwater in farming irrigation areas, it is necessary to ensure the effective replenishment of groundwater through rainfall infiltration.

When the irrigation mode is shifted from traditional flood irrigation to mulched drip irrigation under mulch, mulch on the ground surface changes the path through which water flows between the atmosphere and the soil, and thus exerts an impact on the soil microenvironment (Li *et al.* 2015). After the field water cycle is changed to a minor cycle of water under mulch, the water continuously condenses into water droplets on the inside of the mulch and then drips into the soil. In this way, the limited water in the soil circulates between the soil and the mulch, thus reducing the soil evaporation of crops (Zhao *et al.* 2015). In addition, mulched drip irrigation mainly increases the soil moisture in the root-mixture of the crop. It carries out controlled irrigation at fixed times, fixed locations and with fixed water quantity according to the water demand for the growth of crop. This means that there will be no irrigation return water in mulched drip irrigation. Therefore, compared with traditional flood irrigation, the impact of mulched drip irrigation under mulch upon groundwater is mainly reflected through the following two changes after it is implemented: first, the ground surface mulch changes the conditions of the underlying surface. Due to the impermeability of the mulch, the rainwater falling on the mulch cannot directly penetrate into the soil after it falls on the mulch. It only pools into the bare land (ridge and furrow) between mulches and infiltrates into the soil after it overflows the water interception surface on the mulch. During this process, part of the water is lost and the amount of infiltrated water is thus reduced; second, due to refined irrigation settings, irrigation return water will no longer be generated. This reduces the potential replenishment of groundwater, and thus affects the hydrological cycle process.

Design and measurement

In this paper, South Tallin Aile Village, Yaolin Maodu Town, Horqin Zuoyi Middle Banner, is selected as the observation point, which is located at 44°06′44″N and 122°18′21″E and belongs to the Shebotu Irrigation Area. The soil parameters are shown in Table 1.

HOBO-U30 is adopted to automatically monitor the dynamic situation of soil water content in the section

Table 1 | Soil parameters

Depth (cm)	Soil	Field capacity (%)	Saturated water content (%)
0–30	Sandy loam	26.6	44.8
30–60	Clay	31.2	49.4
60–200	Sandy soil	21.0	28.5
>200	Sandy soil	21.0	28.5

plane. A monitoring section is set up for flood irrigation and mulched drip irrigation under mulch respectively, with 10 probes evenly arranged between 0 to 100 cm and 5 probes between 100 to 200 cm. The maximum monitoring depth for both of the two irrigation methods is 2 meters, and the observation period is from May 24 to August 22, 2017. The data are recorded at an interval of 2 hours. The specific arrangement of the probes for the two irrigation methods is shown in Figure 3. The rainfall data is from the weather station in the monitoring area. As shown in Table 2, a total of 7 rainfalls has been recorded, of which the maximum rainfall was recorded as 72 mm on August 3.

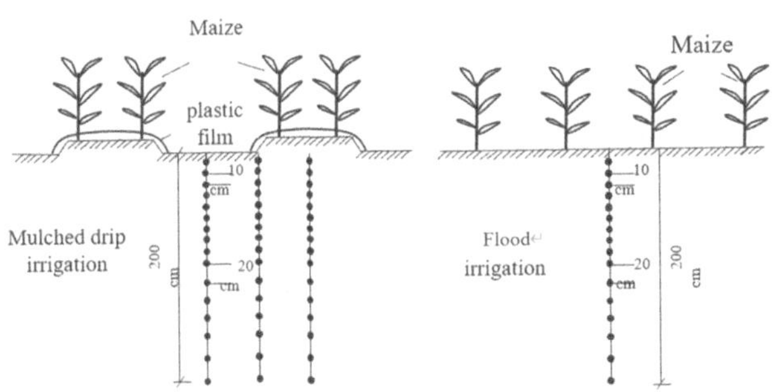

Figure 3 | The arrangement of the probes for the two irrigation methods.

Table 2 | Rainfall information

No.	Date	Rainfall (mm)	Rainfall duration (h)
1	2017/7/7	11.6	3
2	2017/7/9	5.2	3
3	2017/8/3	72	23.5
4	2017/8/10	32.5	10
5	2017/8/11	5.4	1.5
6	2017/8/14	7	1.5
7	2017/8/15	4.2	2.5

RESULTS AND DISCUSSION

Changes in soil moisture content

As shown in Figure 4, in terms of soil moisture content, the flood irrigation area responds more quickly to rainfall than the area of mulched drip irrigation under mulch in general. In the flood irrigation area, each rainfall increases the soil moisture content at the depth of 50 and 100 cm. In the area of mulched drip irrigation under mulch, however, there is no rise in the soil moisture content under the condition of a small rainfall; the soil moisture content at the depth of 50 cm rises rapidly and the soil moisture content at the depth of 100 cm and below rises slowly only when the rainfall reaches 72 mm.

Under the condition of a small rainfall, for instance a rainfall of 11.6 mm on July 7: for the flood irrigation area, the soil moisture content at the depth of 50 cm increased by 3.3% two hours after the rainfall, and the soil moisture content at the depth of 100 cm increased by 2% four hours after the rainfall; in the area of mulched drip irrigation under mulch, however, since the ground surface mulch has changed the condition of the underlying surface, rainwater falling on the mulch cannot directly infiltrate into the soil. It only pools into the ridge and furrow and infiltrates into the soil after it overflows the water interception surface on the mulch. During this rainfall, there was no significant change in the soil moisture content at the depth of 50 cm and below. It shows that when the rainfall is small, flood irrigation is more conducive to rainfall infiltration and replenishment of groundwater than mulched drip irrigation under mulch. Under the condition of a small rainfall, however, the rainfall infiltration depth in the flood irrigation area is less than 1.6 m, which is still within the phreatic limit evaporation depth. If no heavy rainfall occurs in a short period of time, the impact of rainfall infiltration on the water content of the soil aquifer will eventually disappear with the effect of evaporation.

Under the condition of a rainstorm, for instance a rainfall of 72 mm on August 3: in the flood irrigation area, the soil moisture content at the depth of 50 cm increased by 7% two hours after the rainfall, the soil moisture content at the depth of 100 cm increased by 4.1% four hours after the rainfall, and the soil moisture content at the depth of

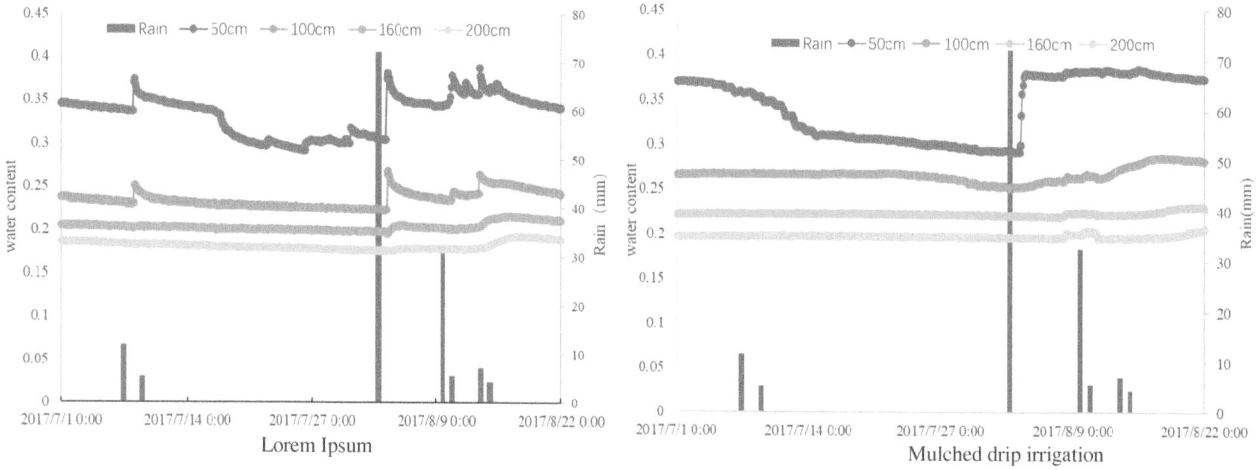

Figure 4 | Effect of precipitation infiltration on soil moisture content.

160 cm increased by 0.9% one day after the rainfall; the changes in the soil moisture content in the area of mulched drip irrigation under mulch, however, lagged behind compared with those in the flood irrigation area. Within one day after the rainfall, only the soil moisture content at the depth of 50 cm gradually increased by 8.3% 12 hours after the rainfall, and the soil moisture content at the depth of 100 cm only increased by 0.1% one day after the rainfall. It is worth noting that after the rainstorm, a rainfall of 32.5 mm occurred in the study area on August 10. As the interval between the two rainfalls was only one week, the soil moisture content at the depth of 2 m increased by 1.4% 14 days after the rainstorm; in the area of mulched drip irrigation under mulch, however, the soil moisture content at the depth of 2 m increased by 0.9% 19 days after the rainstorm, which proves that under the condition of a rainstorm, both the rainfall infiltration depth and the infiltration replenishment amount in the flood irrigation area are greater than those in the area of mulched drip irrigation under mulch.

Changes in groundwater depth

There are few studies on the critical level of groundwater recharge by rainfall infiltration: Zhang (2012) expounded the physical mechanism of infiltration for groundwater supply critical depth, bulit the rainfall infiltration depth calculation model, and the result shows that the general infiltration depth is 7 m to 10 m.

Due to the limited conditions of the research, only soil moisture content within the depth of 2 m were monitored in this study. In the study area, however, the groundwater depth were all over 3 m. In order to analyze the replenishment effect of rainfall infiltration upon groundwater under different irrigation conditions, the monthly groundwater depth average data of seven groundwater monitoring wells in Horqin Left Wing Middle Benner from 2014 to 2018 were collected and compiled in this study, among which four monitoring wells are located in the area of mulched drip irrigation under mulch, namely Wujianfang, Hailijin, Houzhaosileng and Hailiyingzi respectively, and three monitoring wells are located in the flood irrigation area, namely Baiyintala, Mendazhen and Fuyucun respectively (Figure 5).

As shown in Figure 5, in both the flood irrigation area and the area of mulched drip irrigation under mulch, it shows that under the same rainfall conditions, the shallower the groundwater depth, the more severe the impact of rainfall infiltration on the groundwater. As the groundwater depth deepens, the groundwater level rises only when heavy monthly rainfall occurs, and the greater the rainfall is, the earlier the groundwater level rises.

In comparison, more dramatic response to rainfall infiltration and quicker response time for groundwater depth to rise occur in flood irrigation than in mulched drip irrigation under mulch. In the flood irrigation area, the groundwater

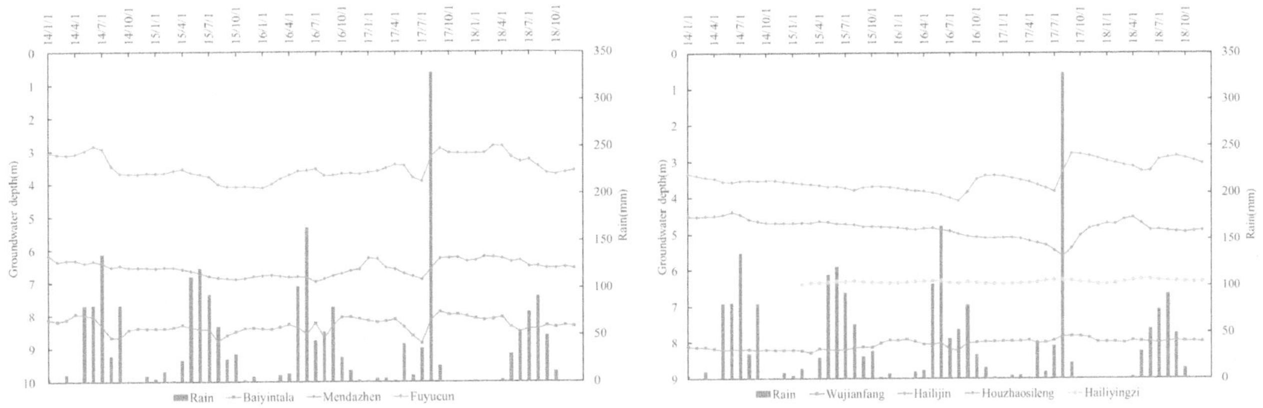

Figure 5 | Effect of precipitation infiltration on groundwater depth. Flood irrigation. Mulched drip irrigation.

depth at about 8 m has already no obvious response to a small amount of rainfall. But when the monthly rainfall exceeds 300 mm, the groundwater depth rises by 0.98 m one month after the rainfall. In the area of mulched drip irrigation under mulch, when the groundwater depth is 4 to 5 m, it rises by 0.77 m three months after the rainfall only when the monthly rainfall exceeds 300 mm; when the groundwater depth is 6 m, there is no obvious response to rainfall.

It is shown that for the West Liaohe Plain, where flood irrigation is a major means of irrigation, the promotion of mulched drip irrigation under mulch will exert a significant impact on rainfall infiltration and the quantity of irrigation return water, and change the critical depth of rainfall infiltration replenishment. On the one hand, there will be basically no irrigation return water generated, and on the other hand, the shift will affect rainfall infiltration and replenishment of groundwater, a key process of the hydrological cycle, both of which will affect the balance of extraction and replenishment of underground water. Therefore, appropriate groundwater extraction levels should be set for flood irrigation and mulched drip irrigation.

CONCLUSIONS

For the West Liaohe Plain, which is a typical semi-arid area, the process of rainfall infiltration to replenish with groundwater is a key link in its vertical hydrological cycle. With different irrigation methods, the replenishment depth of groundwater through rainfall infiltration is different. In this paper, we compare and analyze the impact upon soil moisture movement and water infiltration after the shift of irrigation method from flood irrigation to mulched drip irrigation under mulch through setting up in-field in-situ observation points in the Horqin Left Wing middle Benner in West Liaohe Plain and carrying out groundwater depth dynamic observation. According to monitoring data of soil moisture content, each rainfall in the flood irrigation area increases the soil moisture content. In the area of mulched drip irrigation under mulch, however, there is no rise of the soil moisture content under the condition of a small rainfall; the soil moisture content at the depth of 50 cm rises rapidly and the soil moisture content at the depth of 100 cm and below rises slowly only when the rainfall reaches 72 mm. In terms of the effect of rainfall infiltration to replenish with groundwater, flood irrigation has a stronger response to rainfall infiltration and has a quicker response time in the rise of underground depth compared with mulched drip irrigation under mulch. With the decrease of groundwater level, the effect of rainfall infiltration to replenish with groundwater is significantly weakened. It shows that in the flood irrigation area, the groundwater depth at about 8 m has already no obvious response to a small amount of rainfall. Only when the monthly rainfall exceeds 300 mm, the groundwater depth rises by 0.98 m one month after the rainfall. However, the groundwater depth at 6 m in the area of mulched drip irrigation under mulch has already no response to rainfall. Therefore, when groundwater extraction will be carried out in irrigation areas, reasonable groundwater extraction levels should be designated in light of different irrigation methods to maintain the sustainable utilization of groundwater.

ACKNOWLEDGEMENTS

This research was funded by the Ministry of Water Resources Special Funds for Scientific Research on Public Causes (grant number 201501031), National Natural Science Foundation of China (grant number 41861124006) and National Key Research and Development Plan (No. 2016YFC0503502).

DATA AVAILABILITY STATEMENT

All relevant data are included in the paper or its Supplementary Information.

REFERENCES

Arbat, G., Puig-Bargués, J., Barragán, J., Bonany, J. & Ramírez, F. 2008 Monitoring soil water status for micro-irrigation management versus modelling approach. *Biosystems Engineering* **100** (2), 286–296.

Bufon, V. B., Lascano, R. J., Bednarz, C., Booker, J. D. & Gitz, D. C. 2012 Soil water content on drip irrigated cotton: comparison of measured and simulated values obtained with the Hydrus 2-D model. *Irrigation Science* **30** (4), 259–273.

Huo, S. Y. & Jin, M. G. 2015 Effects of precipitation and irrigation on vertical groundwater recharge. *Hydrogeology & Engineering Geology* **42** (5), 6–13.

Li, X. Y., Shi, H. B., Šimůnek, J., Gong, X. W. & Peng, Z. Y. 2015 Modeling soil water dynamics in a drip-irrigated intercropping field under plastic mulch. *Irrigation Science* **33** (4), 289–302.

Qin, S. J., Li, S., Kang, S. Z., Du, T. S., Tong, L. & Ding, R. S. 2016 Can the drip irrigation under film mulch reduce crop evapotranspiration and save water under the sufficient irrigation condition? *Agricultural Water Management* **177** (C), 128–137.

Rimon, Y., Dahan, O., Nativ, R. & Geyer, S. 2007 Water percolation through the deep vadose zone and groundwater recharge: preliminary results based on a new vadose zone monitoring system. *Water Resources Research* **43** (5), W05402.1–W05402.12.

Shi, L. S., Cai, S. Y. & Yang, J. Z. 2007 Study on spatial variability of subrainfall infiltration coefficient and simulation of the stochastic field. *Journal of Hydraulic Engineering* **38** (1), 79–85.

Skaggs, T. H., Trout, T. J. & Rothfuss, Y. 2012 Drip irrigation water distribution patterns: effects of emitter rate, pulsing, and antecedent water. *Soil Science Society of America Journal* **74** (6), 1886–1896.

Wei, X. M. & Ba, D. D. 2003 Groundwater resource evolution and agricultural eco-environmental issues in the irrigation areas in north China. *Journal of Irrigation and Drainage* **22** (5), 25–28.

Xia, J. & Shi, W. 2016 Perspective on water security issue of changing environment in China. *Journal of Hydraulic Engineering* **47** (3), 292–301.

Yan, L., Chen, M. J. & Ma, J. 2018 A study of the ecological patterns of farming-pastoral zones in semi-arid regions: using the West Liaohe River plain as an example. *Applied Ecology And Environmental Research* **16**, 3229–3244.

Zhang, Q. X. 2012 *Study on the Eco-Hydrological Mechanism in Semi-Arid Regions of China-Taking West Liao River Piain as an Example.*

Zhang, G. H., Fei, Y. H., Shen, J. M. & Yang, L. Z. 2007 Influence of unsaturated zone thickness on precipitation infiltration for recharge of groundwater. *Journal of Hydraulic Engineering* **38** (5), 611–617.

Zhao, X. J., Zhang, Z. T., Zhang, Z. L., Ma, Y. F. & Wu, Z. D. 2015 Soil evaporation for summer maize under no-tillage with straw mulching. *Journal of Drainage and Irrigation Machinery Engineering* **33** (12), 1085–1090.

First received 27 August 2020; accepted in revised form 24 December 2020. Available online 11 January 2021

Calculation of drainage volume during tunnel construction based on the control of negative effects of ecosystem

Yu Wang, Yan Zhao, Xiaoyan Ge, Yongyu Li, Tianrong Huang and Qingming Zhang

ABSTRACT

Water inrush and seepage in tunnel engineering often lead to the loss of regional environmental groundwater resources and the imbalance of the hydrological ecological circulation system, which have a serious impact on the growth of natural vegetation. Based on the analysis of the burial depth of the critical water level of vegetation groundwater, a calculation method for the response of vegetation to water environment stress is established, and the water inrush and seepage volume of tunnel engineering under the condition of ecological balance protection are derived. By calculating the range and volume of the dewatering funnel when the tunnel underground water level reaches the maximum depth after water inrush and seepage, the maximum drainage value that must be controlled in tunnel construction is obtained. The proposed method has achieved certain practical results and could provide some theoretical reference for guiding the reasonable protection of groundwater, alleviating and eliminating adverse environmental geological effects in tunnel construction.

Key words | ecology, environment, groundwater, tunnel, vegetation

Yu Wang
Yan Zhao (corresponding author)
Xiaoyan Ge
Yongyu Li
Tianrong Huang
Shanghai Urban Construction Vocational College,
Shanghai 201415,
China
E-mail: *113120917@qq.com*

Yu Wang
Shanghai Institute of Technology,
Shanghai, 201418,
China

Xiaoyan Ge
Donghua University,
Shanghai 201620,
China

Qingming Zhang
Chongqing Communications Construction (Group) Co., Ltd,
Chongqing 401121,
China

HIGHLIGHTS

- The impacts of tunnel construction on an ecosystem can include chemical and biological impacts on the ecological functions and balance.
- For construction of the tunnel, the reasonable ecological water consumption can be calculated based on the above-mentioned water.
- In order to keep the groundwater level line above the burial depth of the critical water level of the root system, the drawdown of groundwater needs to be calculated.

INTRODUCTION

Mountain-crossing tunnels are usually constructed in complex and diverse groundwater environments, and changes in the groundwater environment will have a direct or indirect impact on the ecosystem and natural vegetation. As a kind of human activity, tunnel engineering often interferes to a certain extent with self-balance and self-circulation of regional groundwater, resulting in different environmental problems (Qumot 1990; Cherry 1998). For instance, lining seepage and long-term drainage in the tunnel often lead to exhaustion of groundwater, and subsequently, drops in the groundwater level, ground surface settlement and collapse, water source cut-off and so forth. In particular, the filling of the seepage channel in the surrounding rock will be gradually taken away by groundwater, which strengthens the channel connectivity and increases losses of water resources, causing damage to the local water environment of vegetation.

doi: 10.2166/ws.2021.012

Therefore, it is necessary to study the impact of tunnel engineering on the ecosystem and vegetation, adopt scientific and effective engineering techniques and plans, and evaluate the impact with scientific methods, so that coordinated development between the ecosystem, vegetation and tunnel engineering can be achieved.

METHODS

Impacts of tunnel engineering on the ecosystem

The impacts of tunnel construction on the ecosystem can be physical, chemical and biological impacts on the ecological functions and balance. As shown in Figure 1, disturbance and imbalance of the hydrologic system are the most obvious and typical embodiments of such impacts, and the degree of disturbance is closely related to the local groundwater content, round-robin manner of connectivity, geological conditions, precipitation, burial depth of the tunnel and construction methods of the tunnel engineering project.

Tunnel construction will inevitably affect the rock and soil structure and cause a stress-field redistribution and failure of the hydrologic round-robin system, changing the interaction mechanism and physical parameters of the material components, affecting the self-balance system of groundwater and resulting in ground surface settlement and collapse, water and soil loss and reduction of surface water, and so on. In areas with complex hydrogeological conditions or water-rich regions, inappropriate tunnel engineering methods (Wei 2009) will disturb the balance of regional water circulation, resulting in a larger range of the dewatering funnel and more ecosystem problems.

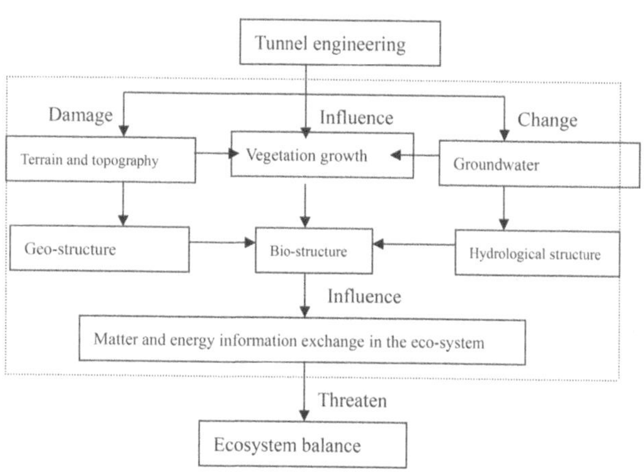

Figure 1 | Impact of tunnel engineering on the ecosystem and correlations between factors.

Research on water requirement of vegetation

In general, the growth and development of vegetation depends on specific soil conditions, climate conditions, groundwater environment and so on. Among them, the groundwater level has direct impact on the development of vegetation. For example, a plant will inevitably wither and fade when the groundwater level is lower than the critical water level of its root system (Voorti 2005).

As shown in Figure 2, the water inrush and seepage due to tunnel construction will cause groundwater level drops and formation of a dewatering funnel in the cave roof aquifer area, forming a new groundwater level line (namely the phreatic line) after groundwater stabilization at the end. After tunnel construction, in order to keep the groundwater level line above the burial depth of the critical water level of root system, the drawdown of groundwater and the range of the dewatering funnel need to be calculated, so as to guarantee the balance between the normal growth of vegetation

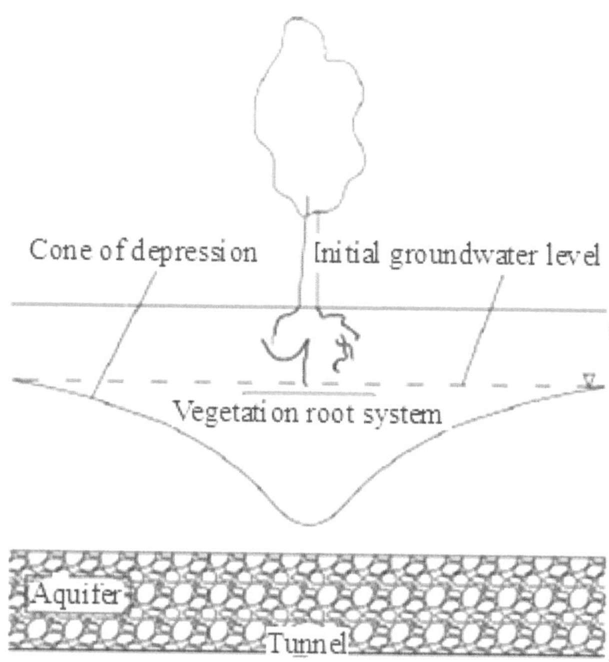

Figure 2 | Impact of tunnel engineering on the groundwater level.

and the regional ecosystem, thus controlling the maximum water yield of the tunnel.

There are three sources of vegetation water supply: (1) natural precipitation; (2) surface runoff and irrigation; (3) groundwater. The rise of capillary water can fully moisturize the soil or make the root system of vegetation directly reach phreatic aquifer when the water table is high, while the plants can grow for a long time.

The water requirement of natural vegetation and ecosystem in a specific region can be calculated on the basis of the amount of the phreatic loss, namely:

$$W = \sum W_i = \sum A_i \cdot \omega_{gi} \cdot K \tag{1}$$

where ω_{gi} is the amount of phreatic loss at a certain burial depth of groundwater level for vegetation type; K is the coefficients of vegetation, and A_i is the area of a certain vegetation type at a certain phreatic position.

The application range of formula 1 is based on groundwater level region. There is a nonlinear relationship between diving loss and plant water demand. Plant water requirements vary with the water level. When the water level is higher, the root system needs more water. When water levels are low, plant roots require less water. If the tunnel displaces more than the ecological carrying capacity, the plants will dehydrate and die.

During the design and construction of the tunnel, the reasonable ecological water consumption in the regional tunnel can be calculated based on the above-mentioned water usage method of vegetation and the ecosystem. Also, through investigation of the local hydrological conditions and precipitation, the water supply of vegetation that needs to be provided by the groundwater can be obtained; and then by back-calculation, the groundwater level required to support normal growth of vegetation can be achieved, which will be taken as the allowable drainage water volume of the tunnel to avoid damages on the ecosystem.

Calculation of water inrush during tunnel engineering

For the tunnel under the incomplete phreatic type of buried condition, the Sato method is usually employed to calculate the maximum water inrush q_0 per unit length, the normal water inrush q_s and the water inrush q_t at the moment t since the maximum water inrush begins to decay:

$$q_{0-0} = \frac{2Km\pi(H - r_0)}{\ln\left[tg\frac{\pi(2H - 3r_0)}{4h_0}\cot\frac{\pi r_0}{4h_0}\right]} \tag{2}$$

$$q_{s-0} = q_{0-0} - 0.584\varepsilon K r_0 \tag{3}$$

$$q_{t-0} = q_0 - \varepsilon\frac{K^2 t}{\lambda B}\cdot\frac{r_0}{q_0}K r_0 \tag{4}$$

where K is the permeability coefficient of the aquifer (m/d); H is the height from static water level to tunnel bottom (m); r_0 is the equivalent circle radius of the tunnel cross-section (m); m is the transformation coefficient, which is set at 0.86; h_c is the thickness of the aquifer (m); ε is the coefficient of experiment, which is set at 12.8; λ is the water-supply degree of the aquifer; B is the tunnel hole width before lining (m); t is the arbitrarily decreasing time between t_0 and t_s (d); and $t_s = \frac{0.584\lambda B q_0}{K^2 r_0}$.

The Ochiai Toshiro, Oshima Yuko and the empirical formula can be used to calculate the tunnel buried depth in the confined aquifer:

$$q_{0-1} = \frac{2K\pi H - r_0)m}{\ln[2(H - r_0)/r_0]} \tag{5}$$

$$q_{0-2} = 0.0255 + 1.9224KH \tag{6}$$

$$q_{s-2} = K\left[\frac{H^2 - h_0^2}{R - B/2} + \frac{\pi H - h_0)}{\ln[4R/B]}\right] \tag{7}$$

$$q_{s-2} = KH(0.676 - 0.06K) \tag{8}$$

In the above formulae, R is the influencing radius of tunnel water inrush (m), and h_0 is the assumed depth of drainage ditch in tunnel cave (m).

The calculation of the maximum water inflow is controlled by many factors. The above empirical formula is also affected by the rock-soil water content, porosity, density and water permeability. In addition, tunnel drainage rate is also a key factor.

Calculation of the cone of depression in the tunnel

During tunnel engineering, the form of groundwater will gradually change with the drainage of groundwater, resulting in a drop in the water level and impacts on the hydraulic gradient and direction of motion. According to the related calculation methods of groundwater, the tunnel seepage can refer to the large-caliber pumping model, and using the well-point dewatering for analysis. After a certain period of constant flow pumping, there will be a cone of depression (Wilcox 2010). Based on the range and volume of the cone of depression and the average burial depth of vegetation groundwater, the allowable maximum drainage volume of the tunnel that satisfies the water requirement of vegetation can be obtained.

Due to the depth of vegetation groundwater balance, it will be affected by root ductility. According to the well point precipitation analysis, the funnel morphology has a close influence on the root depth.

Calculation of the range of the cone of depression

First of all, a clear understanding of the dewatering funnel area is helpful for defining the range of ecological effects. The dewatering funnel area in tunnel projects is always in the form of an inverted elliptic cone (Zhong-xin 2005). And the radial flow of non-standard inverted elliptic cone is similar to the radial flow of a standard inverted cone, which can be expressed by the ground influencing radius R of the cone of depression.

The radius of the dewatering funnel in the aquifer of the tunnel roof is:

$$R_0 = R + \frac{B}{2} \tag{9}$$

Based on the calculation model of the groundwater dewatering radius $R(t)$ under an unsteady flow state and the Theis equation, we obtain:

$$s = \frac{Q}{4T\pi} W(\phi) \tag{10}$$

$$\phi = \frac{r^2 \mu}{4Tt} \tag{11}$$

$$W\phi = \int_0^{\infty} \frac{e^{-y}}{y} dy = \phi - \ln\phi - \sum_2^{\infty} (-1)^n \frac{\phi^n}{n \cdot n!} - 0.58 \tag{12}$$

where r is the distance between the tunnel wall (mm); t is the seepage time (d); Q is the flow amount at the moment (t); s is the decline depth at a certain point in the pumping range at a certain time (d); T is the transmissivity coefficient; $W(\phi)$ is the well function; and μ is the coefficient of water storage.

The largest decline depth of ground influencing radius is generally considered to be 0 mm, namely $s = 0$ mm, and then:

For confined groundwater,

$$R = \frac{3(Tt)^{0.5}}{2\mu^{0.5}} \tag{13}$$

For phreatic groundwater,

$$R = \frac{3(h_c \mathrm{Kt})^{0.5}}{2\lambda^{0.5}} \tag{14}$$

where K is the permeability coefficient of the aquifer, λ is the water-supply degree of the aquifer; h_c is the depth of the drainage ditch in the tunnel cave (m).

This is the important basis for controlling the displacement of the tunnel based on the calculation model of the groundwater dewatering radius $R(t)$ under an unsteady flow state,

Since the water inrush time t in the tunnel is not long before constructing the lining, the ϕ will be larger at a larger distance r from the tunnel. Due to the limitation of the applicability of the above formulae, it is necessary to derive the Theis curve within the range of $0.001 \leq \phi \leq 1$ by the analytical method:

$$W(\phi) = 10.94\phi^{-0.066} - 10.86 \tag{15}$$

And the well function is:

$$W(\phi) = \int_0^{\infty} e^{-\phi} d\phi \tag{16}$$

The decline depth of water level:

$$s = (10.94\phi^{-0.66} - 10.86)\frac{q_t}{4T\pi} \tag{17}$$

For the ground influencing radius of the confined groundwater:

$$R = 2.15(\frac{Tt}{\phi})^{0.5} \tag{18}$$

For the ground influencing radius of the phreatic groundwater:

$$R = 2.15\left(\frac{h_c K t}{\lambda}\right)^{0.5} \tag{19}$$

According to the above formulae, the influencing radius is obtained, so that the boundary of the dewatering area can be drawn.

The long diameter of the longitudinal ellipse boundary of the tunnel is:

$$R_1 = 2R + L \tag{20}$$

The short diameter of the transverse ellipse boundary of the tunnel is:

$$R_2 = 2R + B \tag{21}$$

The calculation of the range of the descending funnel is based on the theoretical formula. Because of the discreteness of parameter acquisition, the radius of the funnel will be affected according to the local conditions.

Calculation of the volume of the cone of depression

The above analysis has clearly defined the range of the cone of depression. Based on this, using the geometric method to analyze the total volume of the cone of depression (V), the total water content in the dewatering area can be calculated to determine the allowable drainage amount of the tunnel. According to the geometric analysis, the volume of the dewatering funnel in the tunnel similar to the inverted cone can be calculated by the following formula:

$$V = \int_0^R 2\pi r s dr \tag{22}$$

Substitute the former formula $s = (10.94\phi^{-0.66} - 10.86)\frac{q_t}{4T\pi}$ and $\phi = \frac{r^2\mu}{4Tt}$ into the Formula (22), and we obtain:

$$V = 3.21\frac{q_t}{T}\left(\frac{Tt}{\mu}\right)^{0.066} R^{1.87} - 2.71\frac{q_t R}{T} \tag{23}$$

Consider the decline depth s_1 of the cave wall and the width B of the tunnel, we achieve:

$$V = 3.21\frac{q_t}{T}\left(\frac{Tt}{\mu}\right)^{0.066} R^{1.87} - 2.71\frac{q_t R}{T} + Bs_1 L \tag{24}$$

At this time, the average decline depth H of the water level within the drainage range is:

$$H = \frac{3V}{\pi R^2} \tag{25}$$

Finally, the total drainage limitation value Q of the groundwater is:

$$Q = V\mu \tag{26}$$

The above analysis can provide some accurate calculation results under certain conditions when the drainage amount is all provided by the raw storage capacity of groundwater in the surrounding rock, consuming the raw storage capacity all the time. The above calculation is particularly suitable for the dry season.

However, since the groundwater system is a complex and connective system, especially in the wet season, the water exchange between the funnel area and the outside regions is very frequent, and the supply increment brought by precipitation restrains the further development of the funnel to some extent (Donoho 1995; Scott 2013). It is necessary to ensure the moisture in the dry season to prevent ecosystem degradation for the water requirement of vegetation. Therefore, the above formulae can be used as an important reference for evaluation.

The numerical relationship between the volume calculation results of the funnel and the displacement of the tunnel is clear, but there is uncertainty caused by the diversity of water supply sources in the wet season and in the dry season.

Groundwater exchange between the tunnel area and the external environment

As the tunnel draining proceeds and the cone of depression develops, the groundwater in the wet season can be effectively supplemented by precipitation and ditch-water supply increment. Meanwhile, there will be declines in natural discharge such as the decrease of spring flow during the median water period and the dry season.

The drainage volume of the tunnel can be expressed as (Rao 2008):

$$Q_d = \mu A(\frac{\Delta h}{\Delta t}) + \Delta Q_r + \Delta Q_d \tag{27}$$

where Δh is the average precipitation level in the drainage affected area within the time period Δt; ΔQ_r is the supply increment during the tunnel drainage period, including the infiltration supply and seepage supply of the precipitation and surface water; ΔQ_d is the decrement of natural discharges during the tunnel drainage period, such as the decrease or cease of the overflow volume of spring water and the evaporation volume of phreatic water.

During the long-term drainage process of the tunnel, if the drainage amount of the tunnel is equal to the sum of the reduction of supply increment and excretion, that is (Zeng-Rong 2001):

$$Q_d = \Delta Q_r + \Delta Q_d \tag{28}$$

And then,

$$\mu A(\frac{\Delta h}{\Delta t}) \rightarrow 0 \tag{29}$$

There must exist: $\Delta h \rightarrow 0$.

At this moment, a new dynamic balance appears in the groundwater system of the surrounding rock of the tunnel: the drainage volume does not rely on the storage capacity any more, and thus the funnel area will not expand further and will gradually enter a stable state. Because some water supply is from spring water, which is a natural and directly usable form of groundwater, the decline or recession of the spring flow will lead to new conflicts in water supply and demand, and result in negative impacts on the environment. In other words, the stable state comes at a cost.

Project-based case study

Introduction to the project

The Huaying mountain tunnel is 4.69 km long. During the construction process, five serious water inrush events occurred in the western section of the tunnel, with a large volume of instantaneous water inrush and eruption of high-pressure water and mud flow, leading the project to be suspended for 69 days. To address this problem, the engineers redesigned the drainage scheme: the radius of the tunnel is set at 2,900 mm, the long draining hole is 1.3 km long, and the radius of the central water drainage pipe for joint drainage is 500 mm. The observation result shows that about 85% of water in the cave in the wet season and about 65% in the median water period and dry season can be drained. Although this scheme presents a remarkable drainage effect, there are still concerns that this will incur abrupt ecological imbalance.

The water inrush section of the Huaying mountain tunnel is about 1,175 m long and passes through the confined aquifer. The lithology mainly belongs to sandstone, siltstone and mudstone. The distance of the confined groundwater is $H = 239m$, the thickness of aquifer is $h_c = 478m$, the coefficient of permeability is $K = 0.0249m/d$, the coefficient of water storage is $\mu = 0.061$, the coefficient of transmissibility in the confined aquifer is $T = 19.05$, the equivalent circle radius of tunnel cross-section is $r_0 = 7.02m$, the tunnel hole width is $B = 12.4m$. The allowable maximum decline depth of tunnel that satisfies the water requirement from vegetation ($s_1 = 28.6m$) is calculated by the research on the regional ecosystem.

Calculation of water inrush

According to Formulae (2)–(4):

The maximum water inrush per unit length is: $q_{0-0} = 8.03m^3/d \cdot m$;

The normal water inrush is: $q_{s-0} = 6.51m^3/d \cdot mq_{t-0} = 7.99m^3/d \cdot m$. It can be found that time has little impact on the water inrush amount and volume, so we assume the groundwater level of the tunnel has reached the maximum decline depth $s_1 = 28.6m$ after $10d$ of excavation.

Then, according to Formulae (5)–(8):

$$q_{0-1} = 7.95m^3/d \cdot m \quad q_{0-2} = 12.41m^3/d \cdot m$$

$$q_{s-1} = 12.04m^3/d \cdot m \quad q_{s-2} = 5.17m^3/d \cdot m$$

Substitute the values of q_{0-1} and q_{0-2} into Formula (4), we obtain:

$$q_{t-1} = 7.93m^3/d \cdot m \quad q_{t-2} = 12.39m^3/d \cdot m$$

The composite average values are obtained:

$$q_0 = 9.48m^3/d \cdot m \quad q_s = 7.86m^3/d \cdot m \quad q_t = 9.45m^3/d \cdot m$$

Calculation of the cone of depression

According to Formulae (19)–(21), we obtain the dewatering radius ($R = 73.8m$), and thereby the long diameter of the top ellipse of the dewatering funnel ($R_1 = 1257.1m$) and the short diameter of the ellipse ($R_2 = 183m$). Due to the application range of Formula $\phi = \frac{r^2\mu}{4Tt}$ is $0.001 \leq \phi \leq 1$, the corresponding influencing range is $2.75m - 83.5m$, that is, the minimum applicable hole distance is $2.75m$ away from the tunnel hole wall, and the calculation width of the tunnel is $B = 12.4 + 2.75 \times 2 = 17.9m$. The volume of standard inverted cone ($V = 601759.32\text{m}^3$) can be obtained by Formulae (22)–(25).

The maximum allowable drainage volume of the tunnel

The result can be calculated by Formula (26):

$$Q = V\mu = 601759.32 \times 0.061 = 36707.30m^3$$

RESULTS AND DISCUSSION

Water inrush and vegetation destruction caused by tunnel engineering have an impact on the structure and functions of the ecosystem. Therefore, in tunnel engineering, we should turn from speed-oriented construction to environmentally friendly construction, resolve conflicts between the ecological carrying capacity and tunnel engineering, and thereby make tunnel engineering more eco-sustainable. Study on the impacts of tunnel engineering on the ecosystem can help understand the internal factors of ecosystem disturbance and imbalance and provide a theoretical basis for tunnel engineering to prevent or reduce its adverse effects on the environment.

CONCLUSIONS

Based on the theoretical methods of groundwater dynamics, we established a quantitative analysis method for the water inrush in tunnel construction and derived the calculation formulae for the volume of the cone of depression after water inrush and seepage. The allowable volume of water inrush and seepage that satisfies the water requirements of vegetation were obtained based on the hydrogeological parameters. The research result therefore can provide a basis for selection of methods to address groundwater-related problems in tunnel engineering.

The groundwater dynamics formulae are derived under certain assumptions, which may deviate from the actual conditions of the tunnel construction projects. In actual tunnel engineering projects, geological conditions and the vegetation are different as the location of the project differs, so it is advisable to adjust these formulae according to the local conditions.

FUNDING

(1) The National Natural Science Foundation of China (51408086).
(2) 2020 Chongqing Transportation Bureau Project.

DATA AVAILABILITY STATEMENT

All relevant data are included in the paper or its Supplementary Information.

REFERENCES

Cherry, J. 1998 *Groundwater*. Earthquake Publishing, Pittsburgh, PA.

Donoho, D. L. 1995 *Ecohydrology Modeling of Lucin*. University of Zululand press, Kwadlangezwa, South Africa.

Qumot, E. 1990 Estimating of normal groundwater distributions. *JASC* **85**, 659–668.

Rao, G. 2008 The treatment of groundwater contaminated data. *Applied Statistics* **24** (2), 218–226.

Scott, C. H. 2013 A review of information on interaction between vegetation and groundwater. *Water SA* **38** (1), 126–139.

Voorti, D. 2005 *Personal Communication Division of Water Environment and Forestry Technology*. Stellenbosch Press, Stockholm, Sweden.

Wei, Z. 2009 Analysis and measures for tunnels water gushing. *Tunnel Construction* **73** (5), 257–261.

Wilcox, B. P. 2010 Ecohydrology of water-limited environment: a scientific vision. *Water Resources Research* **44** (6), 10–24.

Zeng-Rong, X. 2001 Structure depth and relation with fissure water at mountainous regions. *Journal of Railway Engineering* **25** (3), 92–95.

Zhong-xin, J. 2005 Interaction between tunnel and water environment. *Chinese Journal of Rock Mechanics and Engineering* **24** (1), 121–127.

First received 29 August 2020; accepted in revised form 24 December 2020. Available online 12 January 2021

IWA Publishing's authorised EU representative for General Product Safety Regulations is Diane D'Arras, 15 rue Duret, 75116 Paris, France, e-mail: safety@iwap.co.uk.

Printed and bound by CPI Group (UK) Ltd, Croydon, CR0 4YY
12/05/2026
02108874-0004